Hurricane
Almanac

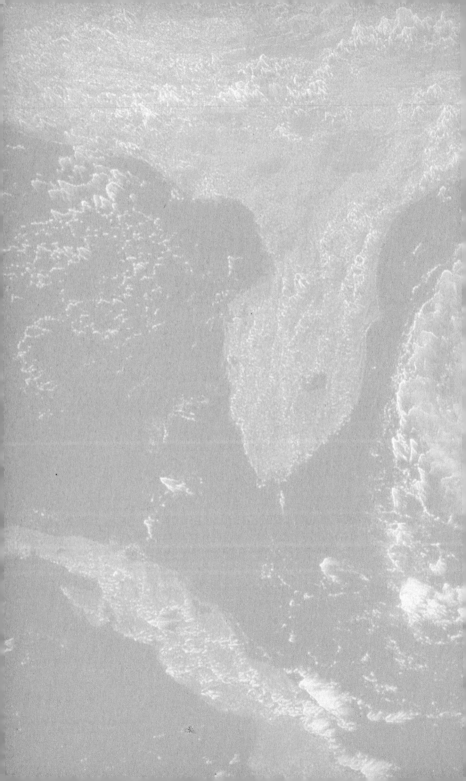

Hurricane Almanac

The Essential Guide to Storms Past, Present, and Future

BRYAN NORCROSS

St. Martin's Griffin ✺ New York

www.stmartins.com

Design by Patrice Sheridan

ISBN-13: 978-0-312-37152-4
ISBN-10: 0-312-37152-7

Revised and Updated Edition: June 2007

10 9 8 7 6 5 4 3 2 1

Max Mayfield doing one of thousands of interviews.

After thirty-four years in government service, Max Mayfield, the intrepid director of the National Hurricane Center, retired in January of 2007. Most people have no idea how hard a job it is these days to run the National Hurricane Center. But, living in Miami and knowing Max and his team at the NHC so well, I've seen it firsthand.

The role of the Hurricane Center director as the spokesman during storms was established by Gordon Dunn in the 1960s. Dr. Neil Frank expanded the job in the 1970s and 1980s to include a role as the most credible advocate for hurricane preparedness. This valuable, but time-consuming, mission was enhanced by Dr. Bob Sheets in the late 1980s and early 1990s and taken to another level again by Max Mayfield.

These days, if there is any downtime, it normally occurs *during* hurricane season. Otherwise, the travel is nonstop, and e-mails and interviews number in the thousands. Through it all, Max has handled the challenges, frustrations, and rewards of leading the nation's hurricane effort with a class and style that will be emulated well into the future.

Best wishes, Max. You've been an inspiration.

Contents

Living Successfully in the Hurricane Zone

Thoughts from Bryan Norcross

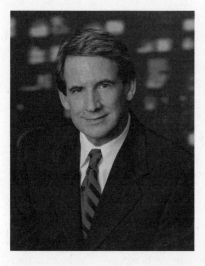

Welcome to the second *Hurricane Almanac*. As I write this in late 2006, the question I hear over and over is "What happened to all of the hurricanes?" Well, El Niño happened. Unexpectedly. As I'll discuss on page 108, the periodic warming of the ocean water in the tropical Pacific produces upper winds unfavorable for hurricane development in the Atlantic. Get the El Niño forecast wrong, and odds are you'll get the hurricane-season forecast wrong as well.

Qualitatively the season behaved somewhat as expected. The storms formed in the deep tropics, east of the Caribbean islands—as opposed to 2005 when the significant ones formed close to the United States—and then curved north. If the Bermuda high had been a few hundred miles farther west, those arcing tracks would have been near the U.S. east coast, and your impression of the "busyness" of the season would be very different.

Quantitatively, the midseason El Niño ended significant storm development early, so the total number of storms was well below what was forecast.

We'll have to watch out in 2007, however. Both 1983 and 1992 were post–El Niño years with few storms, but one major landfalling hurricane. Alicia hit Houston in 1983, and, of course, Andrew hit South Florida in 1992.

We've learned some lessons in the last year. The Hurricanes vs. Global Warming controversy has heated up and, in my mind, become somewhat clearer. There's been more, careful analysis of the data.

The cause of the New Orleans catastrophe is now known, and as I speculated in the previous edition of the almanac, it was man-made.

The full dimensions of Hurricane Wilma have been tallied up, and up, and up. We now know that Wilma was the third most expensive hurricane in U.S. history. Third! From a low-end category 2 hurricane. That should frighten every resident and public official in the hurricane zone. You would think it would have spurred dramatic action in state capitals from Maine to Texas, not to mention Washington, D.C. But, not with the people in charge these days. We've heard hardly a mention at the state level, and nary a peep out of Congress.

And, the coastal-insurance crisis has only worsened. As I'll discuss in "The Insurance Crisis," on page 176, there is only one solution, but, again, the do-nothing Congress lived up to its name.

The companion Web site, www.HurricaneAlmanac.com, will continue to keep you up-to-date with changes, updates, and additions. I want this book—this body of knowledge—to continue to be organic. I've heard from many readers with good suggestions. And I'm sure more will come in, in the year to come.

This year's book divides neatly at page 199. The first part is designed to inform and educate on the state of hurricane science, history, and the processes in place to deal with nature's greatest storms. The second part of the book, called "Living Successfully in the Hurricane Zone," is all about preparation. Along with www.HurricaneAlmanac.com and www.OneStorm.org, you'll find all of the information and resources that I'm aware of to guide you through the steps you need to take to deal with the hurricane threat.

If you live in Michigan, for instance, you wouldn't think twice about building a house with thick insulation and a good heater. If you live in the hurricane zone, you can't debate the wisdom of preparation and planning for hurricanes. Accept it as part of life. Hurricanes happen. The downside of not preparing is too great if the odds catch up with you.

Everyone is, of course, hoping for a quiet hurricane season in 2007. But, the reality is, hoping does not stop the wind and the storm surge and the misery. Only preparation, good sense, and a little bit of good luck can do that. I hope this book and the accompanying Web site will encourage you to take action and painlessly guide you through the necessary steps to safeguard your home and your family.

New in This Edition

Many things have happened in the past year in hurricane world, and that new information is reflected in this book. The sections on Hurricanes Katrina and Wilma, for example, are updated, as well as the rankings of the all-time most expensive hurricanes.

Many people have told me that they enjoyed the section on hurricane history, so that's been expanded by adding some little-talked-about hurricanes that would be catastrophic if they occurred today. The category 4, 1947 storm that hit just north of Fort Lauderdale is an example. When you look at more hurricanes in the aggregate, you see many years when more than one devastating storm occurred. The inescapable meaning is, 2004 and 2005 were not flukes.

Through the last year I've run into some new products and heard more good ideas for living successfully in the hurricane zone—using the dishwasher as a last-resort storage spot for valuables, for example. I've added these ideas and suggestions to the preparedness section.

There are many new sections on the procedures and policies—past and present—of the National Hurricane Center and the National Weather Service. Do you know why *Z-time* uses the letter *Z*, or what a neutercane is? Those sections are among those added.

And I've added to the section somewhat brazenly called "How I'd Do It Better." Many common-sense things could be done to make the hurricane problem at the least less daunting and at the best manageable.

I am hopeful that putting some of the solutions down in black and white might, at least, begin a conversation. The new Congress would seem to be more inclined to do something about the insurance crisis gripping the South, and especially Florida. But, only time will tell. A combination of ineffective government and the natural viscosity of the bureaucracy implies that we may be talking about the same obvious solutions for these critical problems for years to come.

WWW.HURRICANEALMANAC.COM

The free companion Web site has additional resources to keep you informed and to help you prepare for hurricanes, including:

- Additional material and information developed after the almanac was printed
- Hard-to-find hurricane supplies
- Links to Bryan's favorite Web sites for hurricane info
- Frequently asked questions about hurricanes
- *Hurricane Almanac Newsletter* sign-up
- How to order autographed copies of the *Hurricane Almanac*
- Contact info to write to Bryan Norcross

Your questions, comments, and suggestions are encouraged.

1

Hurricanes Today

A New Era of Hurricanes—1995–2006

Hurricanes come in cycles. A casual glance at a list of active hurricane seasons shows clusters of activity in the 1880s–1890s and 1920s–1960s, while the early twentieth century and the 1970s, 1980s, and early 1990s look relatively calm. Another active period started in 1995 and continues today.

Dramatic economic growth, immigration, increasing longevity, and other factors coincided with the hurricane downturn of the late 1960s to mid-1990s, an unfortunate and dangerous confluence of events. The effect can be seen throughout the coastal zone. Would 18 million people be living in Florida today if hurricanes had continued coming at the rate they occurred in the late 1940s and 2004 and 2005? I wonder.

Unfortunately, the people in government with the responsibility for seeing that citizens are safe during disasters were (and are) not good students of history. Most coastal areas, including the state of Florida, were developed without regard to the hurricanes of the past and without protection from the hurricanes to come. Thus, future disasters are guaranteed.

Hurricane cycles correlate with the natural fluctuation in the temperature of the ocean.

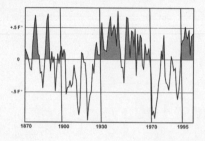

This chart compares the Atlantic Ocean water temperature to normal. The match with hurricane activity is remarkable. Notice that the late 1800s and the period from the late 1920s to the late 1960s show warmer than normal temperatures.

They were active hurricane periods as well. The swing in temperature is only about plus or minus one degree Fahrenheit, but the amount of energy that slight difference makes, when spread out over the entire Atlantic Basin, is significant. See "Global Warming and Hurricanes," page 112.

Hurricane Season 2006

Every hurricane has its own personality, and every hurricane season does as well. The contrast between hurricane activity in 2006 and in the mega year of 2005 could not be sharper. The number of storms was down by about two-thirds, and overall tropical-cyclone activity was down dramatically as well.

Four issues came into play:

- The temperature of the ocean water
- The upper-level winds
- The amount of moisture in the atmosphere over the tropical Atlantic
- Factors we don't understand.

Ocean temperature. The sea-surface temperatures in the tropical Atlantic were somewhat cooler in 2006 than in the superheated year before, but still much warmer than the long-term average. The water was plenty warm for hurricanes to develop.

Upper-level winds. There was a lot of talk about the upper-level winds being unfavorable for hurricane formation, and at times that was the case. But, during the heart of the hurricane season, the wind regime was no less favorable than it was in 2005 over the prime development areas. Some systems were, no doubt, "sheared" by upper-level winds, but that happens to *some* systems every year. Even if some of the upper winds were driven by El Niño, clearly there were other factors involved.

Dry air. The air in the mid and upper levels of the tropical Atlantic was unusually dry in 2006. A tropical system is fueled by moist air from the ocean surface rising into the midlevels of the atmosphere. If the air there is dry, however, the fueling mechanism is disrupted.

The dryness was likely caused by two mechanisms. It seemed that an unusual amount of dry Saharan air lingered over the tropical Atlantic deep into the hurricane season. Also, the wind flow created by El Niño appears to have contributed.

The upper-level winds discussed above operate in a horizontal direction, shearing the tops off systems as they try to develop. There is also a vertical component to the wind, however. The El Niño–driven rising air in the Pacific spreads across the Atlantic, then descends, creating a "circulation cell." The descending air warms *and* dries as it moves lower in the atmosphere where the barometric pressure is higher.

Everyday examples of this double-barreled physical phenomenon:

- A tire pump gets hot because air heats up when it is compressed, everything else being equal.
- A hair dryer warms the air going through it. Warmer air has a greater capacity to hold water. Therefore the air coming out the nozzle acts like a bigger sponge, holding more moisture. The *amount* of water in the air isn't reduced, just its percentage of the air's maximum capacity, making the air feel drier. (This percentage is known as the "relative humidity.")

Factors we don't understand: Even with the somewhat cooler sea-surface, El Niño-affected upper-level winds, and drier-than-normal air, the atmospheric pattern over the hurricane-development areas was not all that bad. There were times during the 2006 hurricane season when the atmosphere appeared primed for tropical development, but it didn't happen. Clearly there are other subtle factors involved that need more study to detect and understand.

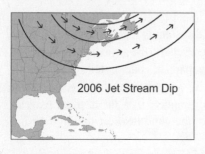

2006 Jet Stream Dip

Jet-stream dip. Another player in the 2006 hurricane season was a persistent dip in the jet stream over the east coast of the United States.

The net effect of this pattern was to drive most of the significant storms to the north toward Bermuda and eventually the Maritimes of eastern Canada. Again, whether this unusually persistent dip was *caused* by El Niño, *exacerbated* by El Niño, or mostly caused by the natural randomness of weather patterns is unclear.

The net effect, however, was that the Bermuda high was much farther east than in 2005, and the steering currents kept the storms well away from the east coast.

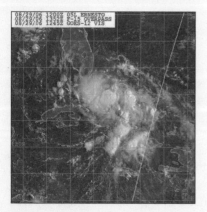

Tropical Storm Ernesto over the Florida straits, August 29, 2006. It appeared conditions were favorable for strengthening, but subtle factors kept the system's top winds at only 45 mph.

Tropical Storm Ernesto. The only significant storm to affect the United States was Tropical Storm Ernesto, which came ashore in the Florida Keys the evening of August 29, 2006. (One year to the day after Katrina hit New Orleans.) It was mostly a nonevent at landfall.

Ernesto's center spent nearly twenty-four hours over the Cuban landmass, which no doubt disrupted the storm's low-level circulation. If it had emerged into the Straits of Florida sooner and had, therefore, twelve more hours over the deep, warm water, it could have been an entirely different story in South Florida. Ernesto continued to strengthen *after* it moved ashore over the Florida peninsula. (The central pressure at landfall was 1004 mb when it came over the Upper Keys, 1,003 mb when it made landfall on the southern tip of the peninsula, and 1,000 mb when the center moved offshore near Cape Canaveral a day later.)

Subjectively, the system had a reasonably well-defined circulation during its time over the Straits, and the water was plenty warm. It appears that some dry air and some unfavorable winds impacted the system during that time; the Hurricane Hunters reported that the system was slightly tilted. The computer models indicated that Ernesto would overcome the factors and be at least a strong tropical storm at landfall, but it was not to be.

The mechanisms involved were subtle. In the end, the computer models and the National Hurricane Center had the right idea. The track was accurately forecast for the sixty hours before landfall, although before that time the forecast cone was well west of the final track. The lesson here, for most people, is that little attention should be paid to the 5-Day forecast cone (I would prefer that the NHC didn't offer it). The 3-Day Cone is the one to pay attention to.

The intensity forecast, on the other hand, never picked up the factors inhibiting intensification. But, based on the best risk analysis that modern science could offer, hurricane warnings were issued for the southern Florida peninsula. There was a good enough chance that Ernesto would intensify that full preparation was required.

Ernesto was a reminder that sometimes—quite often actually—things turn out for the good. Every storm that comes along doesn't blow up in to a mega monster. That fact was easy to forget after the hurricane barrage of 2004 and 2005.

Hurricane Season 2005

Anytime there are a lot of landfalling storms, let alone a record-setting season such as 2005, lessons abound—big and small. In contrast to 2006, the biggest lesson of the hurricane season of 2005 was, the worst does happen!

Hurricane Katrina reaches category 5
strength, August 28, 2005.

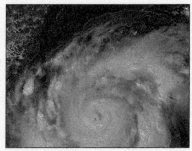

Hurricane Wilma reaches category 5
strength with the lowest pressure ever
measured in the Atlantic Basin,
October 19, 2005. Courtesy NASA.

In case after case people suffered and property was destroyed because someone—a governmental body or official, a business owner or a private citizen—decided to ignore hurricane history, ignore hurricane research, and hope for the best. But, for hour after grueling hour on television we saw that hoping does not hold back wind and water.

The two most remarkable storms of 2005 were Hurricanes Katrina and Wilma. That's not to ignore or discount the effects of Dennis, Emily, Ophelia, Rita, Stan, Gamma, and the rest. In a normal hurricane season, any one of those storms would have been memorable. But Katrina and Wilma rose above the pack in this extraordinary season of storms.

An easy-to-forget aspect of the 2005 season was that none of the record-setting hurricanes formed in the deep tropics. Through history, most of the "great" hurricanes have formed well east of the Lesser Antilles. From there they have time to organize and gain strength. That the traditional breeding ground for big, powerful hurricanes was not fertile in 2005, and yet fifteen hurricanes formed, gives us pause.

Traditional thinking, before 2005, was that a hurricane season when the storms developed mostly on the western side of the Atlantic or in the Caribbean would likely not be as bad because, on average, they wouldn't have as much time to develop. That's still probably a good thought, except when the weather pattern is ideal and the water temperature is like a bath. Those factors came together in 2005.

Hurricane Katrina—Monday, August 29, 2005

Katrina was a hurricane catastrophe . . . in *southern Mississippi*. In Louisiana, it was a healthy hurricane hit . . . and a levee-design/engineering/debacle compounded by a FEMA fiasco . . . that led to human suffering on a scale that we have not seen in the United States in modern times. Everyone involved, from Washington to Louisiana, should be ashamed and embarrassed. We should all be disgusted with our government(s).

Katrina's winds in New Orleans. There is little, if any, reason to think that most of New Orleans experienced more than a category 1 hurricane. For the city it was nowhere near a historic hurricane event. Here is the evidence:

- The only winds in New Orleans measured at over category 1 strength were on the far eastern side of the city. The highest was at the NASA Michoud Assembly plant fifteen miles east of downtown, where the winds were estimated to have gusted to 123 mph. The anemometer was about forty feet high. Another location nearby had a wind gust estimated at 120 mph, giving credence to the idea that these observations were probably accurate. Even using a conservative reduction factor of 20 percent to estimate the *sustained* winds (see page 85) occurring at the same time, we find that it's unlikely that there were values higher than the low end of the category 2 range, if that. And remember, this is in the extreme eastern edge of the city, closest to the circulation center. All of the other measurements—although there weren't many—were low-end category 1 or tropical storm force.
- An anemometer in the middle of Lake Pontchartrain that survived measured a peak sustained wind of 78 mph. While this may not have been the highest wind over the lake, it should be reasonably representative. And it's low-end category 1.
- The water did not top the Lake Pontchartrain levee. Originally designed for a category 3 storm surge, the levee at its existing height would likely only withstand a category 2, according to research

by a group at Louisiana State University. That levee, along with all of southern Louisiana, is sinking. So the evidence is that the storm surge rose to less than category 2 levels.

- The wind rating of the roof shingles on most of the houses in New Orleans was 60 mph. Most of the shingles stayed on.

For these reasons and more it is clear that Hurricane Katrina's winds were *not* the cause of the New Orleans catastrophe.

Katrina's storm surge in East New Orleans. New Orleans is subject to storm surges from two directions from a hurricane moving by to the east of the city. To understand what happened during Katrina, we have to look at them separately. To the east side of the city, a massive storm surge was generated by

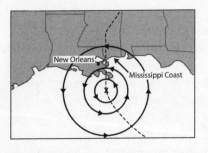

Katrina's winds, which the day before had been 165 mph. That water was forced into the "corner" formed by the southeastern-Louisiana and southern-Mississippi coasts, and then up the Mississippi River–Gulf Outlet Canal (the MR-GO) toward the city. Under the pressure of this water, the levees on both sides of the canal failed, inundating New Orleans East and St. Bernard's Parish. The high-speed water continued west into the Industrial Canal, where the levees also collapsed, putting the Lower Ninth Ward underwater.

The MR-GO Canal had previously been identified as a threat to the eastern part of the city. As water funnels into a narrow canal, the power of the water is accentuated, an extremely bad design in a storm-surge-prone area.

Many areas on the eastern side of New Orleans would have flooded even if the levees had not failed. There is a debate about whether the flooding would have been as bad—it would probably not have been—but it would have been devastating and deadly in any case. It has been well-known for years that this is the most vulnerable part of the city from a storm on Katrina's track. Unless a change is made in the design of the MR-GO Canal, which aims water from the Gulf at the heart of the city, or the channel is closed, the Katrina tragedy in eastern New Orleans will be repeated.

Katrina's storm surge from Lake Pontchartrain. Late in the morning of August 29, Katrina was moving ashore in southern Mississippi. At that time, the northerly winds on the left side of the circulation were pushing the lake water toward New Orleans, into the north-opening drainage canals, and deep into the city. The walls collapsed, with devastating results for the low-lying, northern part of the city.

The flooding should not have happened. We now know that the floodwalls failed because of faulty design and construction by the Army Corp of Engineers. As noted above, the winds pushing the water were likely in the category 1 range. The water pressure appears to have been well below the levees' advertised specifications.

The whole thing just looks like a bad idea. These north-south canals—like the MR-GO Canal to the east—allow the powerful storm surge deep into the city. Strong floodgates should have been put at the entrance to the canals so the floodwalls wouldn't be subject to the extreme pressures. A version of that idea is under way now as a temporary fix.

In addition, big pumps take water out of the city and push it into these canals and *toward* the lake—*against* a storm surge driven by a north wind. How could this ever work? The pumps should dump the water directly into the lake, so the pumped water is not trying to flow against the water pushed by the wind.

Amazingly, the potential of a wall failure on these north-south canals was never brought up in any seminar or conference I ever attended at which the New Orleans hurricane problem was discussed. Far more explanation is going to be required from the experts as to how this could have been overlooked.

Katrina in Mississippi. In Mississippi, Katrina was a historic hurricane. The storm surge exceeded the high-water mark from Hurricane Camille in 1969, the previous benchmark storm for that region. While Katrina was nowhere near as strong as Camille, it was a much bigger storm in physical size. Storms with a large "radius of maximum winds" generate a larger storm surge. See "Storm Surge," page 117.

Any hurricane that produces a twenty-to-thirty-foot storm surge is going to do major damage at the coast. But at least some of the blame needs to fall on the officials in Mississippi who allowed such expensive infrastructure to be built where it was guaranteed to be ruined by any significant storm. When it came time to authorize casino gambling on the coast, Hurricane Camille apparently escaped everybody's memory. It's another example of government creating a policy that's built on wishing and hoping. Government should not roll the dice. We can only hope that, once and for all, this will be enough to indelibly make the point: The worst does happen.

Mississippi building codes. In 2006, Mississippi finally got a building code, sort of. It's not required in much of the state. Local governments are allowed to opt out, for some reason unfathomable to me. Because a city or county commission gets co-opted by outside forces, should the citizens be forced to suffer?

But, having said that, it's a start. In Alabama they haven't even started. It's the only state without a state building code in the most hurricane-vulnerable region, and it's a disgrace.

The resistance to legislation that would protect lives is, in my opinion, impossible to justify. Mississippi has not previously had state standards to protect people from fire, and there are more fire deaths there than in any other state by far. The whole thing is a sad commentary on the state of government.

Florida's Building Code

Florida has had a statewide building code since the late 1990s. By any objective measure the standard in effect in most of the state is pathetic, given the hurricane threat. Shutters are not even required (although there is an effort under way to eliminate that loophole). Only Miami-Dade and Broward counties in southeast Florida were able to hold off the lobbyist-compromised, inadequate code in effect elsewhere in the state. Legislators will tell you, "It was the best we could do." Translation: Too many wrongheaded politicians in Tallahassee were paying more attention to lobbyists than to the safety of their constituents.

When people are injured because a poorly constructed house collapses, rescue personnel often have to risk their lives to help. When buildings come apart and spray debris all over the neighborhood, damage is multiplied and public money is spent to clean up the mess. A strong building code is in the public interest, both in terms of public safety and fiscal responsibility.

FEMA and Michael Brown. In my opinion, fired FEMA director Michael Brown did not get a fair shake. His position and power in the hierarchy of the newly created Department of Homeland Security–FEMA structure didn't give him the authority that the position previously had. He was in a bureaucratic box.

Still, a more courageous and more experienced emergency manager might have walked up to a microphone the day that Katrina hit and pointed a finger at the director of Homeland Security and the president and said, "People are going to die here if you don't do something now!" If he was as frustrated by the bureaucratic bungling as he says he was, why wasn't he pointing with all ten fingers at once and screaming at the top of his lungs? Instead, he played the good soldier and tried to work within a dysfunctional system—that he had been a part of for years.

The former FEMA. FEMA was a model agency before it was folded into the Department of Homeland Security in 2003. The move, per se, wasn't the problem. If the agency had been left intact to do its job, the structural change would not have been disruptive. But people without emergency-management experience made decisions that gutted FEMA's ability to quickly and properly respond to natural disasters. DHS is, essentially, a police agency, cops looking for bad guys. FEMA and emergency management are different animals. In large measure that was and is the problem.

Government disconnect at the highest levels. That President George W. Bush, DHS Secretary Michael Chertoff, and the head of the U.S. Army Corps of Engineers, General Carl Strock, were completely misinformed and saying ridiculous things for days and weeks after the disaster is frightening. These people know when a pin drops in Afghanistan. How can they not know when a levee breaks in New Orleans? The evidence says that the communications and operational infrastructure of the federal government broke down. We should all be very concerned.

Hurricane Wilma—Monday, October 24, 2005

Wilma in the Caribbean. At about 8:00 A.M. EDT on October 19, Air Force Reserve Hurricane Hunters measured the lowest barometric pressure ever seen in the Atlantic Basin (the Atlantic Ocean, the Caribbean Sea, and the Gulf of Mexico): 882 millibars beat the 888 mb measured in Hurricane Gilbert in 1988. And the pressure was still falling. It may have dropped even lower after the measurement was taken. The National Hurricane Center estimates that Wilma reached its peak intensity at about the same time, 185 mph.

In the twenty-four hours after 2:00 A.M. EDT on October 18 Wilma went from a 70 mph tropical storm to a 175 mph, category 5 hurricane. The extraordinary wind speeds were possible because Wilma developed an eye that was only about two miles across. In general, the smaller the eye, the stronger the winds. The physical principle is called conservation of angular momentum. It's the same thing that causes a figure skater to spin faster by pulling his or her arms in toward the body. This rate of strengthening had never before been seen in the Atlantic Basin. In fact, no other storm in the record book had ever strengthened at a rate even close to Wilma's. Also, the staff at the National Hurricane Center noted that they were not aware of an eye ever being seen as small as two miles. When I read the observation from the aircraft, I thought it was a mistake.

Wilma in South Florida.
After the storm bashed the Yucatán Peninsula, the National Hurricane Center forecast showed a track across South Florida. That's when a lot of amateur meteorologists in metropolitan southeast Florida decided (a) a storm coming from the Gulf can't be too bad, and (b) we never get two storms in one year and we've already had Katrina. Wrong and wrong again. It was a painful, aggravating mess for hundreds of thousands of people from Palm Beach County to the Florida Keys—and an extreme inconvenience for the rest of the over 5 million residents on the southeast coast.

Hurricane Wilma was the *third* most expensive hurricane in U.S. history, as of the totals available in late 2006. Yet a good part of the metropolitan Miami–Fort Lauderdale–West Palm Beach area never saw winds higher than category 1. In the National Hurricane Center's analysis, some areas did experience category 2 winds, but most of the category 2 winds appear to have been in gusts. Because of the dense population right along the coast, the effects were magnified. Thus, the second big lesson from 2005:

ANY HURRICANE MOVING OVER A MAJOR METROPOLITAN AREA WILL CAUSE PROBLEMS THAT WILL EXCEED THE ABILITY OF THE GOVERNMENT TO MANAGE THEM.

Government response to Wilma. You hear politicians today patting themselves on the back on how well they did in a "very difficult situation." If a "barely category 2" storm is a very difficult situation, what would that make a category 3 or 4? "Unmanageable with the current system" is the only answer.

In a major city there are too many people, with too many difficult problems for the current underfunded, underdeveloped emergency-management system to handle. FEMA's emasculation has only made the situation worse. The numbers tell the story. If government could solve the major problems of 90 percent of the population in three days (which would never happen, but let's just imagine), more than 500,000 people in metropolitan Miami–Fort Lauderdale–West Palm Beach would be abandoned! And these are the people who likely have no transportation and are the least able to take care of themselves. Hundreds of thousands of people could be hurt and lost. It's a monstrous public safety issue.

The entire system needs to be revamped. Mickey Mouse emergency-management systems with learn-as-you-go politicians making critical decisions on the fly need to be scrapped and professionalized. Politicians who show up and take charge at the time of an emergency make bad decisions because (a) they have not taken the time to be trained in emergency management, where, theoretically, all imaginable scenarios are thought out ahead of time, and (b) their instant decisions on what to say and do are often driven by political (translate: wishful) thinking.

Building code flaws revealed. Throughout this book you'll see references to the mighty South Florida Building Code in use in Miami-Dade and Broward counties. It is a strong code, the strongest in the country against hurricanes. But Wilma pried it open and showed its weaknesses.

First, roofs blew off, and all kinds of bad things happened, mostly in buildings built before the code. Some affordable system of retro-fitting existing structures is essential, or a category 3 hurricane is going to be cataclysmic.

Second, numerous high-rises built to the current code suffered severe damage. Nobody expected that to happen. One flaw in the code is the idea that "large missiles" (big hunks of debris) are only a threat on the lowest three floors of a structure. Now we know better. Because of the way winds swirl in every direction around high-rises, the entire building needs to be resistant to debris impacts, not just the lowest thirty feet. Also, damage to one high-rise sprays debris on other high-rises downstream, aggravating the situation.

Correcting this flaw is an expensive and frustrating reality that will affect every new building. At least Miami-Dade and Broward counties' governments are interested in fixing the problems and doing it right, which is more than you can say for the rest of Florida and most of the hurricane coast.

See "How I'd Do It Better," page 175.

Hurricane Dennis on July 9, 2005, heading toward the areas still trying to recover from 2004's Hurricane Ivan. Courtesy NASA.

Hurricane Dennis—Sunday, July 10, 2005

Dennis's records. Never before in the record book had a hurricane as strong as Dennis formed before August. (See "Emily," page 16.) On July 8, just before making landfall in south-central Cuba, the maximum sustained winds peaked at 150 mph—high-end category 4 strength. After weakening over Cuba, Dennis restrengthened dramatically over the Gulf of Mexico. On the morning of the tenth the maximum winds were up to 145 mph as the storm headed toward the Pensacola-Mobile area. Fortunately, Dennis weakened right before coming ashore, so the estimated maximum winds were 120 mph at landfall, but no observations higher than 100 mph were reported.

Dennis's storm surge. Dennis came ashore about thirty miles east of where Ivan hit in 2004, but was not as strong or as damaging. Still, the storm surge in the western Florida Panhandle was 6 to 7 feet, enough to wash over the barrier islands near Pensacola and do significant damage, especially in Navarre Beach. The surprise, however, occurred in Apalachee Bay, 180 miles to the east. A 6 to 9 foot storm surge there swamped the town of St. Marks and the surrounding area.

The surge was approximately double what was expected. A phenomenon called a "trapped shelf wave" was responsible. Basically, in the shallow water above the very wide continental shelf along the west coast of Florida energy propagated north. Even though the shelf was on the edge of the storm, the "ducting" caused by the shelf had an enhancing effect.

Hurricane Emily—Monday, July 18, 2005

Late on Saturday, July 16, Hurricane Hunters found category 5 winds in Hurricane Emily when the center was about 115 miles southwest of Jamaica. It was a very close call for that island. This was the first and only category 5 hurricane ever seen in July. As with Dennis, the extremely warm Caribbean and Gulf water seems to have been a factor in creating these July storms of record strength. A category 3 version of Emily came ashore over Cozumel on Mexico's Yucatán Peninsula early on the twentieth, an area that would be devastated by Hurricane Wilma three months later.

Hurricane Rita—Landfall: Saturday, September 24, 2005

Southeast Texas and southwestern Louisiana got the short end of the hurricane stick in 2005. When Rita veered north from Houston, the consensus was that "it could have been a whole lot worse." Well, that's certainly not the sentiment in Port Arthur, Texas; Lake Charles, Louisiana, and their environs. A 15-foot storm surge was measured in Cameron, Louisiana, and water pushed thirty miles inland as far as I-10. Towns along and just north of the Gulf in southwest Louisiana were essentially destroyed. The coastal area there may never be the same. Winds were estimated at 115 mph along the coast at landfall, but weakened quickly inland.

Rita in the Keys. The Florida Keys were on high alert for Hurricane Rita. On September 19 Rita was located in the central Bahamas moving toward the Keys and forecast to become a hurricane, perhaps an intense hurricane. Thankfully the center of Rita threaded the needle between the Keys and Cuba keeping the strongest winds offshore, and, more importantly, the expected intensification didn't happen as fast as forecast.

Rita didn't reach hurricane strength until early on September 20 well south of the Upper Keys. By mid-afternoon, when the storm made its closest approach to Key West, the winds had increased to 100 mph, but the center was far enough south to keep the strongest winds off the island. In the next 24 hours Rita intensified into a category 5, 170 mph hurricane. The Keys were extraordinarily lucky that the intensification process held off. If it had begun even 12 hours sooner the effect on the Lower Keys would have been dramatically different. As it was the water rose about 5 feet on the southern side of Key West as Rita went by causing moderate flooding.

Rita in the Gulf. Rita's spectacular rapid intensification phase was similar to what we had seen with Katrina a month before and would see with Wilma a month later. At 11:00 P.M. EDT/10:00 P.M. CDT on the twenty-first, Rita had maximum sustained winds esti- mate of 180 mph and a central pressure of 895 mb. That was the third lowest pressure ever mea- sured in the Atlantic Basin (now it's the fourth).

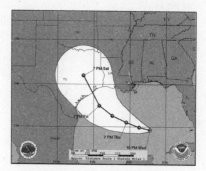

At that time, the forecast cone predicted a landfall somewhere between Corpus Christi and cen- tral Louisiana. The focus was on the Houston-Galveston area, how- ever, because of the potential for a

Hurricane Rita's forecast cone from September 21, 2005.

cataclysmic direct hit there. In fact, the landfall odds were about equal over the entire northern half of the Texas coast, and not significantly lower in western Louisiana.

Houston evacuates. In the Houston metropolitan area it was clear the hurricane plans in place were inadequate and unrealistic. Millions of people tried to evacuate at the same time. It was chaos. You might have heard that the problem was caused or exacerbated by people who did not live in the evacuation zone leaving when "they didn't have to." Well, yes and no.

The problem is that most Houston-area homes don't have shutters and the structures are of questionable strength even if they do. Should people stay in Houston-quality homes in the face of a category 3 or 4 hurricane? My advice would be "Get out!"

The only solution in metropolitan Houston is a crash program to build safe rooms within existing homes. Governments there have allowed this massive metropolitan area to be built with substandard homes, given the hurricane threat. There's no going back and rebuilding correctly. But people can strengthen a closet or a small room so they will be safe at home during a storm.

Hurricane Stan—Tuesday, October 4, 2005

Disastrous flooding occurred in Central America at the time Stan made landfall with 80 mph winds ninety miles south of Veracruz, Mexico. Between 1,500 and 2,000 people are reported to have died, most of them in Guatemala. It is not believed that Hurricane Stan was directly responsible for this event. A large low-pressure system to the east of Stan seems to have to been the cause of the flooding.

Records Set in 2005

MOST NAMED STORMS: 28. PREVIOUS RECORD: 21, in 1933.

MOST HURRICANES: 15. PREVIOUS RECORD: 12, in 1969.

MOST CATEGORY 5 HURRICANES: 4. PREVIOUS RECORD: 2, in 1960 and 1961.

ONLY CATEGORY 5 HURRICANE EVER SEEN IN JULY: Emily, on July 16.

MOST CATEGORY 3–5 U.S. LANDFALLS: 4. PREVIOUS RECORD: 3, in 2004.

MOST NAMED STORMS BEFORE AUGUST 1: 7. PREVIOUS RECORD: 5, in 1997.

COSTLIEST ATLANTIC HURRICANE SEASON: $107+ billion. PREVIOUS RECORD: $45 billion, in 2004.

COSTLIEST HURRICANE: Katrina, at $80+ billion. PREVIOUS RECORD: Andrew, $26.5 billion in 1992 dollars.

DEADLIEST U.S. HURRICANE SINCE 1928, THIRD DEADLIEST EVER: Katrina 1,200+.

LOWEST PRESSURE EVER MEASURED IN THE ATLANTIC BASIN: Wilma 882 mb. PREVIOUS RECORD: Gilbert, 888 mb, in 1988.

GREATEST PRESSURE DROP IN 6 HOURS: Wilma 54 mb. PREVIOUS RECORD: Beulah, 38 mb, in 1967.

GREATEST PRESSURE DROP IN 12 HOURS: Wilma 83 mb. PREVIOUS RECORD: Allen, 48 mb, in 1980.

GREATEST PRESSURE DROP IN 24 HOURS: Wilma 97 mb. PREVIOUS RECORD: Gilbert, 72 mb, in 1988.

MOST TOP 10 LOWEST PRESSURES ON RECORD: Wilma 882 mb (no. 1), Rita 897 mb (no. 4), Katrina 902 mb (no. 6).

Predicting How Busy a Season Will Be

Dr. Bill Gray at Colorado State University pioneered the idea that the amount of tropical storm and hurricane activity during the *forthcoming* season could be predicted. Before Gray issued his first seasonal forecast in 1984, it had not escaped anyone's attention that some hurricane seasons were busy and some were relatively quiet. The reason for this was elusive, however. Each season seemed like a roll of the dice.

Gray identified atmospheric and oceanographic features that were relatively stable and correlated them to hurricane activity. An untold number of measurements of the atmosphere can be taken at any given time. Some, such as the temperature at a certain spot on the earth, are always changing. Others, such as the temperature of the ocean water, change much more slowly. (See "El Niño and La Niña," page 108.) If quantities that are relatively stable for months could be found, they could be compared to hurricane activity to determine if there was a correlation. If a trend in a certain quantity matched a trend in tropical development, it might be used as a "predictor."

The set of predictors that Dr. Gray and his team use to make the seasonal forecast has evolved over the years as new weather patterns have come along. An atmospheric process that was stable in the non-hurricane-friendly pattern of the 1980s might vary radically during the current active period, for example. As new predictors are found, they are tested on hurricane seasons and their attendant weather patterns of the past to be sure they are valid over the varied scenarios. If predictors work for a large number of past seasons, the theory goes, they should work in the future as well.

Beginning with the 2006 seasonal forecasts, Dr. Gray has turned over lead responsibility for the projections to one of his students, Philip Klotzbach, who has been involved with the program for the past five years.

The Predictors

The Colorado State University forecasting team led by Philip Klotzbach and Dr. Bill Gray predicted a busy hurricane season in 2006.

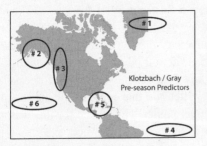

Klotzbach / Gray
Pre-season Predictors

Six predictors of future hurricane activity used by the Klotzbach/Gray team at CSU.

Net Tropical Cyclone Activity, a measure of the busyness of the season, was expected to be about double the normal level in the Atlantic Basin. Instead it was dramatically less than normal. The overriding reason: El Niño. As noted on the next page, El Niño is one of the two dominant predictors that can trump other favorable conditions.

Interestingly, all of the predictors used to make the preseason forecast were favorable for an active season in 2006. Often this is not the case. In an average season, for example, some predictors will point toward more storms, others to fewer storms, and some will be neutral.

The Klotzbach/Gray preseason predictors are:

1. *Midlevel (500mb) atmospheric pressure in the extreme North Atlantic.* When the pressure is high, the ocean is warmer and conditions are normally more favorable over the tropics and the Caribbean for storm development.
2. *Surface pressure in the Gulf of Alaska.* When the pressure in the Gulf of Alaska is low, that normally correlates with the

development of a La Niña event. La Niñas correlate with increased hurricane activity.

3. *Midlevel (500 mb) atmospheric pressure in western North America.* It's been found that high pressure readings in this area during the previous fall often correlate with favorable wind conditions in the upper atmosphere over the tropics during hurricane season.

4. *Upper-level (200 mb) winds over the tropics.* The upper-level wind pattern often changes direction each hurricane season. When the wind is *against* the storms as they move to the west, it's a negative factor.

5. *Surface pressure in the Gulf and Southeast.* When the pressure values are low during the previous fall, pressures often stay low into the hurricane season. Low pressures give storms a head start.

6. *El Niño and La Niña.* The challenge is to forecast whether either condition will exist, or the pattern will be neutral. See page 108.

Two Dominating Factors?

There is some evidence that two factors are the most significant in controlling how many storms develop in a season: the water temperature of the Atlantic Basin and whether an El Niño or La Niña is under way. Many of the Klotzbach/Gray predictors are related to one of those two factors.

Water temperatures in the Atlantic Basin (north of the equator) go up and down in twenty- to forty-year cycles. El Niños and La Niñas occur in three- to seven-year cycles, although there is considerable variability. These cycles on top of cycles account for some of the apparent randomness of the atmospheric-oceanographic system.

Seasonal Forecast Timetable

Updated forecasts for the 2007 season are issued throughout the year by the Klotzbach/Gray Colorado State team. Look for them at www.HurricaneAlmanac.com.

National Oceanic and Atmospheric Administration (NOAA) researchers release their forecasts in mid-May and early August. They will also be on www.HurricaneAlmanac.com.

Hurricane Hot Spots

New York City

The biggest potential hurricane disaster on the Atlantic Coast is in the New York City, northern New Jersey, western Long Island metropolitan area. Hurricanes don't happen often here, and the most populated part of the area has been essentially hurricane-free since the beginning of the twentieth century. But, the nineteenth century was a different story. Numerous storms affected New York City during that time.

It appears the strongest hurricane on record hit New York City on September 3, 1821, possibly as a category 3. The evidence is that the center went over Brooklyn, Queens, and western Connecticut. Severe damage was done in the region and Lower Manhattan was deep under water. (See "Cape May to Long Island Hurricane," page 37, for more.)

Other hurricanes hit the New York City area on June 5, 1825, July 19, 1850 (along with two other tropical storms that year), and August 24, 1893. If central and eastern Long Island are included, the list is much longer.

The threat from hurricanes in the New York metropolitan area is

greater today than ever. More people live near the coast with no organized evacuation plan. The upside-down L-shape of the coastline will act like a funnel feeding water into New York Harbor if a hurricane pushes storm surge toward the corner of the L. The resulting flood could take out the

NYC subway, highway tunnels, and bridges, depending on the storm's intensity. Tall buildings would likely stand, but many would be blown out. Falling and flying glass would be a major threat. There is no apparent solution to the problem in the works.

Miami-Fort Lauderdale

Metropolitan southeast Florida is the American urban center most likely to get hit by a hurricane in any one year. Since Miami was first developed at the turn of the twentieth century, numerous significant hurricanes have made direct hits on the region, but the list has important gaps that have made South Florida extremely vulnerable to big hurricane disasters today.

Significant hurricanes hit the area in 1906, 1926, 1928, 1929, 1935, 1945, 1947 (twice), 1948 (twice), 1949, 1950, 1960, 1964, 1965, 1992, and 2005. This list doesn't count the major hurricanes that hit the Keys early in the century or areas just to the north in 2004.

Notice the two glaring gaps in the list: 1906 to 1926 and 1965 to 1992. Both of these periods saw tremendous development in the region. The giant 1926 hurricane cleaned out the shoddy construction of that era. But small-diameter Hurricane Andrew was not large enough to do the same in 1992. (Thankfully.) Major flaws in the building code, and especially code enforcement, were laid bare in the southern suburbs of Miami by Andrew. But homes built under the same system in northern Miami-Dade County and Broward County (home to Fort Lauderdale) were not affected.

From the early 1970s into the 1990s, as suburban sprawl took over South Florida, construction standards were relaxed. Hurricane amnesia set in at the same time. Roofs were not tied together well and were no longer designed for high wind; shutters were no longer standard equipment; and entry doors and garage doors were weak and vulnerable. Many tens

of thousands of these homes and buildings still stand today, vulnerable to the next major hurricane.

Miami-Dade and Broward counties now have the best building code in the United States for hurricanes. It's not perfect, as Hurricane Wilma proved. But there is a will in South Florida to make it right. Unfortunately, it is not retroactive to existing buildings, so the threat of a mega disaster continues. See "How I'd Do It Better," page 175.

Charleston, South Carolina

If any American city could be called disaster prone, it would be Charleston. In 1698 a smallpox epidemic swept the area, then, in early 1699, there was an earthquake and fire that destroyed a good part of the city, followed by a hurricane that caused a massive flood. The earthquake-hurricane double punch came along again at the end of the nineteenth century. In 1885, what was likely a category 2 hurricane came up from the south and did significant damage. Before rebuilding could be completed, the strongest eastern U.S. earthquake

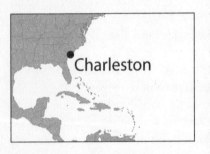

ever hit on August 31, 1886. (The historic buildings of the city still have earthquake bolts holding their walls together.) Then, seven years later, the strongest hurricane ever to hit the region (up to that time) came ashore, killing upward of 2,000 people (see "Sea Islands Hurricane," page 41).

The coastal region of South Carolina is called the Low Country. Obviously, low country and hurricanes are a bad combination. The Carolina coastal islands are extremely vulnerable to being submerged by a storm surge, as they were in 1893 and 1989. And computer projections show the potential for a surge of 20 feet or more in the city.

Evacuation plans are well developed and normally well implemented in the Charleston area, likely because of more than three hundred years of dealing with hurricanes, and a cultural appreciation of history.

New Orleans

The Lower Ninth Ward was underwater. People climbed on their roofs hoping to be rescued; some died trapped in their attics. It was late on the night of September 9, 1965. Hurricane Betsy was passing New Orleans to the west.

Greater New Orleans is, has been, and always will be vulnerable to hurricanes. Even if the levees are raised, made stronger, and, for a change, maintained, there will still be the potential for big problems. In addition, many people in southeastern Louisiana live *outside* the biggest levees. That will always be the case. The reality is that no practical levee can be built to stop every imaginable hurricane. Obviously, a lot can be done with money. Tall, strong levees with big pumps *might* reduce the risk to an acceptable level, but there's another big issue.

Southern Louisiana is sinking. The subsidence rate is about two inches a decade. With the rising sea level, in a hundred years New Orleans is expected to be more than three feet farther below sea level than it is today. A variety of factors contribute to the problem. The levees, highways, and railroad tracks that have blocked the natural processes by which soil is replenished in the Mississippi Delta are likely the biggest culprits.

New Orleans will likely come back from Katrina as a smaller but still vibrant city. If Mother Nature's needs for rebuilding and maintaining the ecosystem of southeastern Louisiana aren't put at the top of the priority list, however, the city and the region won't survive into the indefinite future.

Tampa–St. Petersburg

Most people in Tampa probably don't know it, but they need to be looking hard at Providence, Rhode Island. Providence sits at the top of Narragansett Bay. Up

until 1966, hurricanes that tracked just to the west of the city—and therefore put strong southerly winds over the bay—caused severe flooding downtown. After destructive floods from the Great New England Hurricane of 1938 and Hurricane Carol in 1954, local residents decided to do something. In 1966 the Fox Point Hurricane Barrier was finished. Now, when a hurricane approaches from the south, the barrier is closed and the city is protected.

Tampa Bay looks a lot like Narragansett Bay; it opens to the south. A hurricane on a north-northwest track just to the west of Tampa–St. Petersburg will push a tremendous amount of water into the north end of the bay and inundate much of the area. This worst-case scenario has happened at least four times that we know of.

On September 25, 1848, a hurricane pushed a fifteen-foot storm surge up Tampa Bay. There was not much of a town there at the time, but the event was well documented by a major at Fort Brooke, the military outpost. (See David Ludlum's *Early American Hurricanes, 1492–1870*.) An event like this today would be catastrophic, flooding vast sections of the metropolitan area near the water. Another hurricane came two weeks later, around October 11, 1848, and caused flooding as well, but the surge was only about ten feet. Even that would be tremendously damaging today.

In 1921, a 125 mph hurricane pushed water ten to twelve feet high into the bay. And in 1950, Hurricane Easy was well west, so that winds in Tampa were only 60 mph, but still the water rose ten feet, with waves on top of the surge.

Though hurricanes are not as likely in the Tampa Bay area as they are in Miami, the potential for loss of life is greater on the west coast. Safe areas are, at most, a few miles from the beaches of metropolitan South Florida. But, vast portions of the Tampa–St. Petersburg area are only a few feet above sea level, subject to much higher storm surge, and safety lies on high ground many miles away over low causeways.

2

Hurricane History

Notable Hurricanes

The criteria for designating a hurricane as "notable" are subjective, of course. Perhaps 5,000 storms have come and gone since Christopher Columbus and his crew became the first Europeans (that we know of) to experience a hurricane in 1495. Of those, innumerable hurricanes have had a lasting impact due to the damage done, casualties caused, treasure lost, and military campaigns thwarted. What follows are the stories of nearly fifty important or unusual hurricanes of the last 750 years.

Please refer to page 79 concerning hurricane records. It's important to remember that words like *only* and *strongest* refer to those storms documented in the historical record. There is every possibility that other storms were stronger and more damaging, perhaps even in the time frame covered here, than these storms. We just don't know about them, or the details are sketchy.

Hurricane Andrew
Landfall: Monday, August 24, 1992
Hurricane Intensity: August 22–26

I put Hurricane Andrew first because of the important role the storm played in my life, and the important role it should have played in the development of a permanent, efficient emergency-response system in the United States.

Andrew barely survives. Hurricane Andrew almost didn't make it. The weather pattern was not at all conducive for tropical development that year. The 1991 El Niño lasted until early in the year, so systems were getting blown apart from the strong upper-level winds. Short-lived depressions formed in both June and July in generally unfavorable conditions for development. Then, on August 16, a third tropical depression formed in the far eastern Atlantic and became Tropical Storm Andrew the next day.

On Wednesday, August 19, the satellite presentation was poor; there was little or no surface circulation. I said on the TV news that evening that "Andrew has about a fifty-fifty chance to survive, maybe a little more likely that it will hang together. And if it does, we'll have to watch it carefully." Even then it was clear that high pressure might block the system from turning north, although nobody imagined the kind of hurricane it would become.

The system did, of course, survive, and it steadily strengthened over the next two days. On August 21, Tropical Storm Andrew was 650 miles due east of Miami with 60 mph winds. The morning forecasts from the National Hurricane Center showed the storm bending to the north in the general direction of central Florida, perhaps arriving early the next week. Not a certainty, of course, but we were not very concerned in the southern part of the state.

Friday, August 21. Computer models at that time arrived in our weather office on the slowest printerlike apparatus you could ever imagine. That was the state of the technology in 1992. That Friday afternoon I waited for the (500 mb) forecast charts covering the next three days to come off the machine. When they finally arrived, I could see that the high-pressure system to our north was not just going to stay in place; it was forecast to strengthen and move east over the path of the storm. This meant to me that there was *some* chance that Andrew would stay south, and that *could* mean trouble for us. A stronger high-pressure system also meant the possibility of the hurricane arriving at the coast sooner than the current forecast called for. This was, admittedly, a subjective sense of the possibilities presented by the future weather pattern forecasted by the AVN model, the only model available to us at that time.

I didn't *know* Andrew was coming. I didn't *forecast* Andrew to come. But, I made a judgment, based on the *possible* future weather pattern, that the *risk* was high enough that we needed to raise the possibility that it would happen. I was especially concerned because it was Friday afternoon and many people don't stay in touch with the news over the weekend. So at 3:30 P.M. I went to Sharon Scott, who was in charge of the news at WTVJ (the NBC station in Miami where I worked at the time), and told her we needed to "raise the flag" beginning on the 5:00 P.M. news. I wanted people to know that there was a chance that we might have to deal with Hurricane Andrew over the weekend.

The late afternoon advisory did not cause any big concern. The new official forecast kept the storm reasonably far away from South Florida. The other television stations more or less conveyed that message. By the 11:00 P.M. news the possibility of Andrew affecting South Florida was more widely discussed on the various broadcasts. We were using an early version of the forecast cone graphic on the air, which more clearly showed the chance of the storm's coming our way. But, of course, it also showed the possibility of Andrew veering well north.

Saturday, August 22. The next morning things started to move quickly. Not only did Andrew become a hurricane, it soon had an eye and was picking up forward speed. I raced into the TV station and went on the air at 11:00 A.M. We did hour-long blocks of unscheduled programming during the day as new information came in, and we answered viewers' questions. Hours and hours of viewers' questions. A hurricane watch was issued for the southern half of the east coast of Florida at 5:00 P.M., and we did extended coverage through the evening.

By the 11:00 P.M. news that Saturday night, Andrew was a category 3 hurricane with 125 mph winds and strengthening. At 1:00 A.M., when

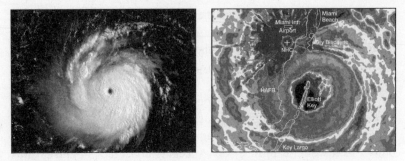

Hurricane Andrew weakened as it moved toward Florida *(left)* but then restrengthened just as it was coming ashore. This is the last full sweep of the Miami National Weather Service radar (Courtesy NASA) *(right)* before it blew off its stand and crashed onto the roof of the building housing the National Hurricane center, August 24, 1992. Courtesy NOAA.

we finally decided to end the newscast, I said, "I'm going to go home now and get some sleep and I suggest you do, too. Tomorrow is going to be a very big day, and I'm not sure if we're going to get sleep tomorrow night," as best I can remember it. So I went home, took care of some things, and went to bed at 2:30 A.M.

Sunday, August 23. At 6:00 A.M. I got up, gave some friends instructions on how to put up the shutters and secure things, took one last look at the house and the neighborhood, and headed for work. From 8:00 to 9:00 A.M. we at the station talked about how to keep news crews safe, where they should go, and some other coverage details.

At 9:00 A.M. that Sunday, August 23, I sat down at the anchor desk along with our sports anchor Tony Segreto, who was raised in South Florida, and news anchor Kelly Craig. We watched it happen. That afternoon Andrew smashed Eleuthera in the eastern Bahamas. Now we know the top winds at the time were about 170 mph. Our coverage was a combination of reporters in the field, viewers' questions, and conversations with the indomitable Dr. Bob Sheets, the director of the National Hurricane Center.

Bob and his staff deserved the Congressional Gold Medal for their performance during Andrew. While this frightening buzzsaw of a hurricane was heading for their homes and families, they put out advisories warning the Gulf Coast of the coming storm. They never missed a step.

It was an incredible display of dedication. Bob was clear and precise and confident through the day and night. He was an inspiration.

I looked at a tape of a good part of the coverage some years later and realized I said some dumb things, but luckily they weren't important. I also said some really good things that came to me as the situation was getting worse and worse. The most important was inspired by a book I had read by L. F. Reardon about his experiences in the

Bryan Norcross, Kelly Craig, and Tony Segreto broadcasting from a storage area off the main studio in the early-morning hours of August 24, 1992. Viewers called in as their houses came apart. Our colleague George Butch took the picture.

1926 Great Miami Hurricane. At one point, when the wind was whistling through his house, he put his children in the laundry washtub and *covered them with a mattress* (smart move) while he and his wife went to the garage to wait out the storm in their car (not so smart).

Early Monday, August 24. When it became evident in the early morning hours of August 24, 1992, that people in southern Dade County were going to have to do everything possible to protect themselves in their homes, I remembered the mattress. I said something like, "Friends, here's what I want you to do. Get a mattress off the bed and have it ready. When you go to your safe spot, get your family in there, get the mattress over them, and wait this thing out." It was the smartest thing I have ever said in my life. The stories I have heard from people who spent the storm under a mattress still give me the chills.

We watched the last sweep live on TV as the radar was blown off the roof of the National Hurricane Center just as the eyewall was coming ashore. Luckily, we had a backup phone line to the National Weather Service radar in West Palm Beach, so we could still track the storm. We were the only station with that access. Our weekend weathercaster, Brien Allen, kept it connected and fed us information throughout the storm. Frightened people were calling in as their homes were coming apart. Everyone just wanted to know when it was going to be over.

The storm was moving fast, about 20 mph. So the worst of it was there and gone in about three and a half hours. The first reaction of the people in South Dade was stunned disbelief. When they opened their doors or got out from under their mattresses, they found it impossible to comprehend what had happened to their neighborhoods and their lives. From the TV studio in downtown Miami, we knew there was damage, but we couldn't tell how much. There was no communication with the south end of the county. I said on the air that the fact we had not heard from Homestead, Cutler Ridge, and points south was *not* good news.

The nation is misinformed. The national news programs that morning reported that Miami had "dodged a bullet." That was just the first unsettling parallel to Hurricane Katrina. The national media were uninformed, but they would not simply say, "We don't know what happened yet." That practice continues today. It was pervasive throughout the 2005 hurricane season. Reporters and meteorologists who make definitive statements about things that are uncertain or unknown are dangerous and do a disservice to their audience.

The "hurricane zone," as it came to be called, extended from Homestead and Florida City in the south up to SW 120th Street in the Kendall section of South Dade County. Not that there wasn't damage farther north, but as you drove down the Dixie Highway (once the debris and light poles were cleared) there was a line below which the devastation was nearly total. If you were not there to see it, you cannot imagine the dimensions of what happened. I had seen total destruction before, after Hurricane Camille, but never on this scale. The destruction did not just extend as far as you could see, but as far as you could see from a helicopter!

Presidential photo op. President George H. W. Bush came to take a look the next day. He promised to help, and promptly went back to Washington. Nothing happened. It was eerie seeing his son do essentially the same thing after Katrina.

Anarchy reigns. By Wednesday the twenty-sixth the situation was worse than I ever imagined it could be in the United States of America.

There was anarchy as perhaps a hundred thousand people were left with no power, no water, no food, and no security. Lots of good neighbors were trying to help, but the relief effort was an uncoordinated nightmare. There were no communications. The governor couldn't even call his office. Every police agency that had radios was on a different frequency. Then Kate Hale, the embattled director of Dade County Emergency Management, went on TV and said, "Where in the hell is the cavalry on this one? They keep saying we're going to get supplies. For God's sake, where are they?"

The military takes over. Somehow that broke the political logjam and the U.S. military was sent to rescue South Dade. That next day, Thursday, August 27, when the first C-5A landed at Opa-Locka, we carried it live on TV. It was the first moment of hope that we'd had that week. I got misty on the set watching that first big plane full of supplies and help. It didn't happen immediately, but by Saturday the military was up and running and assuming control of the situation. A full six days after Andrew hit, people in South Dade got their first night's sleep free from the fear that gangs of looters were going to break into their dark houses.

Andrew's lessons. The lessons of Hurricane Andrew *should have been*:

- Journalists, and especially broadcasters, should never make definitive statements about uncertain things, whether it's the storm's track or the status of neighborhoods, areas, or levees that they haven't seen or know about.
- A presidential visit (or flyover) is a waste of time. Presidents should stay in Washington, get the federal machinery moving, and come visit after they've done their job.
- Key public safety agencies must share radio frequencies. There is no hope of having a coordinated response if responders can't talk to each other.
- Well-built and protected homes survive even category 5 hurricanes. Bob Sheets and I saw many examples of totally destroyed neighborhoods nearby other areas where the better-built homes were still habitable.

And most important:

- The *only* entity in our society that can bring command and control to a catastrophe zone is the U.S. military. They come with housing, transportation, and communications. Any catastrophe plan that doesn't have the military on the ground with a general in charge of organization and security immediately after the event is not a catastrophe plan.

How these lessons could have been lost on the federal bureaucracy in 2005 is just beginning to emerge. I know that many of the good people at FEMA learned the lessons of Andrew and took them to heart. But the fact that the current federal disaster management team was apparently unaware (based on their actions) of the important lessons and events of the most expensive natural disaster in the history of the United States—more than three times any previous disaster—is nothing short of frightening.

Kamikaze Hurricane
Landfall: August 14, 1281

Japan is saved from invasion. The mighty Mongol warrior Kublai Khan had conquered China and Korea by 1274 and had his sights set on Japan. That year he sent a fleet of hundreds of boats to attack, but the Japanese had built a six- to nine-foot-high stone barrier more than twelve miles long (part of it still exists today) for protection. The Battle of Hakata Bay ended when a storm came up and the Mongols retreated. The Khan sent a much larger invasion fleet in 1281 with perhaps 150,000 men. The force made it ashore, but retreated to the ships for safety when a storm approached. The storm turned into a violent hurricane (*typhoon* in today's language) that destroyed the armada and killed most of the men. The Japanese were saved from the Mongol invaders by what they called the "divine wind." The word in Japanese is *kamikaze*. This story is well-known in Japan, although some historians have expressed doubts about some of the details.

Still, the best evidence of its truth might be that the strongest force in the world, Kublai Khan's army, never did conquer Japan.

Columbus's Fourth-Voyage Hurricane
Landfall: Tuesday, July 1, 1502

The first hurricane warning. Christopher Columbus's fourth voyage to the New World brought him to Santo Domingo to replace one of his damaged ships. The sky was milky with fast-moving cirrus clouds, and there was a strong southeast swell. Columbus knew from his previous voyages that these signs foretold a violent storm. Seeing a large fleet in the harbor preparing to sail, he sent a message to his archenemy, the governor of the city, warning of the imminent storm and asking for permission to obtain a new ship. Governor Nicholas de Ovando denied the request and sent the ships on their way toward Spain while Columbus sailed his small fleet to another harbor for safety. Most of the flotilla was destroyed in a hurricane near Puerto Rico that killed five hundred or more men and sent untold treasure to the bottom of the sea. Columbus's fleet was damaged, but survived.

The Great Colonial Hurricane
Landfall: Saturday, August 25, 1635

The first hurricane to hit the American colonies. We know of this hurricane from the writings of John Winthrop of the Massachusetts Bay Colony and William Bradford of Plymouth Plantation. Mention was made of the storm passing the settlement at Jamestown, Virginia, on August 24, but without reference to damage. Correlating with Winthrop's report of a strong northeast wind at Boston and Bradford's of a monstrous storm surge on the south shore of Cape Cod, the apparent track of the storm was near Norfolk, then speeding through New England just to the east of Boston. The storm surge heights would indicate that this was likely a major hurricane.

The Great Hurricane
Barbados Landfall: Tuesday, October 10, 1780

The deadliest Atlantic hurricane on record. Arriving first at Barbados on October 10, this tremendous storm leveled the island in, reportedly, a two-day assault. The great storm lumbered to the west, lashing the Leeward Islands of St. Vincent, St. Lucia, Martinique, St. Eustatius, and Grenada, among others. Thousands of people were killed on these islands, but many more died on the British and French warships that were vying for naval supremacy in the region, adjunctive to the American Revolution. About 3,000 French sailors and soldiers died in Martinique alone. The estimated death toll in the Caribbean was 22,000. The Great Hurricane was just one of eight significant storms that season; two other hurricanes that same month each killed more than 1,000 people. The 1780s was a decade of hyper-Atlantic-hurricane activity in which an estimated 31,000 people died.

The Great September Gale
Landfall: Saturday, September 23, 1815

The most powerful New England hurricane in 180 years causes widespread damage. The storm came ashore on the south shore of Long Island, New York, with great intensity, moving northward, at perhaps 50 mph, through eastern Connecticut and Massachusetts, between Worcester and Boston, and into New Hampshire. On the Charles River in Boston, "spray was raised to the height of 60 to 100 feet." "Immense" damage was reported in Amherst, New Hampshire, near Manchester. But the worst of the damage was in Providence, Rhode Island, as it would be again in 1938, when Narragansett Bay inundated the city. A post rider traveling from west of Worcester to Boston wrote, "There appeared to be one continued scene of devastation, in the unroofing of houses, upsetting of barns, sheds, and other buildings, and in the general prostration of fences, trees, grain, and every description of vegetation" (from David Ludlum's *Early American Hurricanes*).

Cape May to Long Island Hurricane
Landfall: Monday, September 3, 1821

The storm that kick-started hurricane science. The evidence shows that the center of the storm came ashore in eastern North Carolina and rapidly moved north over Norfolk, the Delmarva Peninsula, coastal New Jersey, and New York City and on to New England. The storm devastated Norfolk—in some quarters the storm is called the *Norfolk* to Long Island Hurricane—on the morning of the September 3, inundating the city with a storm surge of at least 10 feet. By early afternoon the eye moved over Cape May, inflicting major damage. The center of the storm appears to have stayed just west of the shoreline as it moved north through New Jersey then over Lower New York Bay which feeds into New York Harbor. Thirteen feet of water was pushed over Lower Manhattan, flooding most of the area up to Canal Street. The lasting legacy of the great 1821 storm is William C. Redfield's close examination of the event, which offered the first concrete evidence and explanation of the structure of a hurricane. If the holiday had been celebrated at the time, this would have been a Labor Day hurricane.

Last Island Hurricane
Landfall: Sunday, August 10, 1856

Summer partygoers "danced until they died." In the mid-nineteenth century, those Louisiana residents who could afford it would vacation on the barrier islands along the southern coast of the state to get away from the mosquitoes and the oppressive heat. It was thought that the mosquitoes that caused the yellow fever epidemic in 1853 couldn't breed in the salt water near the coast. Last Island, on the southern edge of Terrebonne Parish, was a favorite. The small settlement of Village Bayou sported two hotels, a handful of summer cabins, and miles of woods and beaches for hunting and family fun.

On Saturday, August 9, 1856, the weather started to deteriorate. By the next morning the storm was at full roar—the Gulf water sweeping away the town and its inhabitants that afternoon. Today it is believed that a 120 mph, category 3 hurricane tracked over Last Island, moving from southeast to northwest. The estimate is that about 140 of the approximately 400 people on the island died. Many were swept far north into the marshes of southern Louisiana.

The name Last Island translates into French as Île Derniere. The hurricane and coastal erosion have broken the old island into many parts, so today you see it labeled on maps Îsles Dernieres, the plural. *Îsles* is the Old French spelling.

Another major hurricane tracked to the north in the eastern Gulf three weeks later. Weakening slightly, this storm came ashore at or near Panama City as a strong category 2, spreading strong winds and damage as far west as Pensacola and north through southern Alabama, Georgia, and into South Carolina. Because of the widespread destruction, it is known as the **Southeastern States Hurricane**.

Indianola Hurricane
Landfall: Friday, August 20, 1886

Two strikes and you're out. Mother Nature's mood in 1886 appears to have been similar to the nasty temper we experienced in 2005. The storm tracks were concentrated near Florida, the Gulf, and the east coast of the United States. At least seven tropical systems passed over Cuba, and four storms hit Texas. A total of twelve storms are known to have formed that year, with seven making landfall in the United States. But remember, there was no way to know if there was a storm in the tropical Atlantic east of the Lesser Antilles unless a ship reported it, so it's likely the actual number of storms was higher.

From the early 1850s, a military road ran from West Texas through Austin and San Antonio ending at the town of Indianola. This made the town an important port for the military and commerce in the western Gulf, 115 miles southwest of Houston. The location seemed ideal, on the south shore of protected Matagorda Bay. But, on September 16, 1875, a category 3 hurricane swept in from the southeast, pushing a massive storm surge over the Matagorda Peninsula and into Indianola, destroying most of the town.

A smaller version of the town was rebuilt, but the dream of Indianola's growing into a major port wasn't to be. On Thursday afternoon, August 19, 1886, the wind was increasing from the northeast. A major hurricane was in the western Gulf, and it was intensifying. By sunup the next morning top winds had increased to 155 mph, by modern reckoning, with the center thirty-five miles south of Indianola, in a perfect location to push a massive storm surge into Matagorda Bay. The Signal Building, where the weather station was located, collapsed in the morning hours, killing the official observer and starting a fire that spread through part of the town. Buildings that didn't burn were smashed by the violent wind and the surging water.

Indianola succumbed to the category 4 hurricane and was never rebuilt. The county seat was moved to Port Lavaca, 15 miles to the northwest. The estimated lowest central pressure of 925 mb makes the Indianola hurricane of 1886 the ninth strongest on record, and the strongest hurricane in the Atlantic Basin from 1851, when the official record begins, until the confirmed category 5 San Filipe Hurricane that hit Puerto Rico in 1928. See pages 40 and 137. Another, weaker hurricane—this time a category 1—tracked over Corpus Christi and just to the west of Indianola five weeks later, sealing the town's fate.

Seven U.S. landfalls in one season is still the record. The 1886 hurricane season featured two category 1s, three category 2s, and one category 3 and one category 4.

New York City Hurricane
Landfall: Thursday, August 24, 1893

The last hurricane to make a direct hit on New York City. The bad weather began Saturday evening, August 19, 1893, in the New York City region. For eighteen hours the tristate area from New Jersey to Connecticut was pounded by torrential rain and strong northeast winds. The rain came down at 2 to 4 inches per hour in parts of New Jersey. This weather was the edge effect of a category 1 hurricane that passed just east of Long Island on August 21 while, apparently, a cold front moved in from the northwest. The combination caused widespread flooding and temperatures in the sixties.

At the same time, another hurricane located north of Puerto Rico was beginning to arc to the north on a similar path. The New York Weather

Bureau forecaster had warned of another coming storm, but late in the evening of August 23 the wind and rain arrived ahead of schedule. At that time the storm's center was likely located in the Atlantic east of Washington, D.C., but it was speeding to the north and would come ashore near Coney Island on the south side of Brooklyn around sunrise.

According to *The New York Times*, thirty-foot waves crashed ashore, destroying buildings, ripping up railroad tracks, and pushing boats a half mile inland. Central Park "was devastated as never before." Trees were uprooted, and the "drives" (the East River Drive and the Hudson River Drive on either side of Manhattan) were washed away in places by the raging rivers. Winds are now estimated to have been 85 to 90 mph. It's now been 114 years since a hurricane's eye has crossed the city of New York.

Hog Island or *NOT* Hog Island

Many of the accounts of the 1893 New York City hurricane include descriptions of the destruction of a resort on an island called Far Rockaway Beach east of Coney Island and just offshore of Far Rockaway, Queens. Indeed, *The New York Times* tells the story of a family that was rescued from a cabin on the island just before a wave swept it away. The newspaper account does not report that the resort was destroyed, but Professor Nicholas Coch of Queens College and his students discovered artifacts and debris from the period when the U.S. Army Corps of Engineers dredged the nearby ocean bottom to replenish the beach at Far Rockaway.

The island's name is, however, often given as Hog Island. I believe this to be an error. The maps from the late nineteenth and early twentieth centuries at the New York Public Library show the inlet to the east of the island in question as Hog Island Inlet. The island of the same name is shown as being well to the northeast of that inlet. The island with the resort is clearly labeled as Far Rockaway Beach with a causeway linking it to Far Rockaway on the mainland.

Between 1903 and 1906, according to the dates on the maps, Far Rockaway Beach—which was never much more than a sandbar—disappeared completely.

Sea Islands (Georgia and South Carolina) Hurricane
Landfall: Sunday, August 27, 1893

Officially, the third deadliest U.S. hurricane. Three days—if you can imagine—after the New York City hurricane, a significantly stronger storm with top winds at landfall of about 115 mph plowed into the Georgia coast near Savannah. The right side of the storm pushed a massive storm surge over the low-lying Sea Islands near the Georgia–South Carolina border. Most of the residents of the islands at the time were poor former slaves and descendants of slaves. More than 2,000—Clara Barton of the Red Cross estimated the number to be 4,000 to 5,000—drowned as the water swept in around midnight. Those who survived managed to climb trees, the only "high ground" available. The cabins and shanty homes were carried away by the salt water, which ruined the crops and the freshwater supply. Jobs on the plantations and phosphate mines were wiped out by the storm as well. The Red Cross's massive ten-month-long relief effort to stabilize the region was slowed in October when another category 3 hurricane came close before slamming the coast just to the north, near Myrtle Beach.

Chenier Caminada (Louisiana) Hurricane
Landfall: Monday, October 2, 1893

The fourth deadliest U.S. hurricane and the strongest of a very busy year. Exactly a month after the catastrophic events in coastal South Carolina and Georgia, a relatively small hurricane was detected in the northwest Caribbean. Similar in size and track to Hurricane Camille seventy-six years later, it moved north over the extremely warm Gulf waters and came ashore on the barrier islands fifty miles south of New Orleans in the early morning hours of October 2. Winds estimated at 135 mph pushed a giant storm surge over the southeast Louisiana coastal plain and destroyed the town of Chenier Caminada. Remains can be found today on the island just west of Grand Isle on the Gulf Coast. Half of the town's population of 1,500 was killed and a total of 2,000 died—mostly in

Louisiana and Mississippi—as the storm swept northeast, bypassing New Orleans. Only four houses survived in Chenier Caminada and the town was never rebuilt. Survivors moved north to the area near Cut Off, Louisiana, where today they are protected by a levee. During the hurricane season of 1886 more hurricanes hit the United States, but never before or since has there been a season with more than one storm that killed over 1,000 people.

Great Galveston Hurricane
Landfall: Saturday, September 8, 1900

America's deadliest hurricane. For two days beginning on September 3 a weak tropical storm slogged over the eastern two-thirds of Cuba, finally emerging into the Florida Straits just east of Havana on the fifth. As the storm intensified into a category 1 hurricane, the Weather Bureau in Washington forecast the storm to turn north and northeast, threatening the east coast of the United States. But when strong southeast swells started flooding the Galveston, Texas, waterfront early on the September 8 while the wind were coming from the opposite direction, it should have been clear that something was up. Galveston's Weather Bureau chief, Isaac Cline, kept Washington apprised of the developments, but forecasters there never issued a hurricane warning.

By noon that Saturday the wind was howling, and it increased steadily though the afternoon. The winds reached 120 mph or more in the early evening and pushed the Gulf waters high into the city. Cline watched from his house as the water rose twenty feet in a few hours. His wife was killed in the flood, along with untold thousands more. The official death toll is 8,000, by far the deadliest disaster in the history of the United States; most people think the real number was higher. Later analysis showed that storm likely came ashore with 145 mph winds, category 4 on the Saffir-Simpson scale.

At the time of the hurricane, Galveston was the second largest port on the Gulf (next to New Orleans), and much more important than Houston or Dallas. The city was destroyed and never recovered its former prominence.

Florida Keys Hurricane
Landfall: Thursday, October 18, 1906

Miami's first significant hurricane. Ten years after the city of Miami was founded, tropical reality finally arrived. Relatively minor storms had flirted with the fledgling city during four of the years since 1896, but they only reinforced the idea that hurricanes were not a serious problem. The first decade of the new century was a heady time. Henry Flagler's Florida East Coast Railway was bringing the elite from the Northeast to the posh Royal Palm Hotel in the heart of the young city. The future seemed boundless.

To the south, Flagler's workmen were undertaking the seemingly impossible task of extending the railroad "across the ocean" to Key West. In the fall of 1906 hundreds of workers were in the Upper Florida Keys building a roadbed, laying track, and trying to survive in the stiflingly hot, mosquito-filled air. They lived in floating dormitories—large shanties built on top of barges—that could be pulled along as the project progressed.

On October 17, 1906, a compact, fast-moving hurricane moved over western Cuba heading northeast toward Florida and the Keys. The Weather Bureau raised storm warnings in Key West and Miami late that morning, but not everyone got the word. Supervisors at the railroad work camps had makeshift barometers. They saw that the pressure was dropping and alerted the camps to prepare for a storm, but nobody was ready for what was to happen that night.

After dark the wind started to blow at a mighty roar. The quarter boats tugged at their moorings and eventually ripped free, carrying hundreds of men into roiling Florida Bay to the north of the Keys. Some men clung to floating pieces of the demolished dormitories; many others were swept away, never to be found. The exact number of men lost is unknown, but it's believed to be about 160.

Seventy-five miles to the northeast in Miami, the wind started to howl in the early-morning hours of October 18, blowing "a northeast gale . . . with much rain." For thirty minutes beginning about 9:30 A.M. the winds went calm as the eye of the hurricane passed directly over the city—the first of at least six times that would happen in the twentieth century.

Nascent Miami didn't rate a weather station at that time, but modern estimates are that the top winds were about 115 mph. The telegraph office downtown was flooded and streets were impassable from debris and pieces of damaged buildings—a small sample of what was to come when a much bigger hurricane would hit a much bigger Miami head-on just twenty years later.

The railroad project in the Keys was suspended for a year while crews and equipment were reassembled. Thought was given to stopping where they were, in the Middle Keys, but Henry Flagler wouldn't have it. On January 22, 1912, the first scheduled train finally pulled into Key West, but not before construction was interrupted twice more by major hurricanes in 1909 and 1910. While both storms were as strong or stronger than the disastrous 1906 event, neither caused the same level of disruption.

Track of 1910 Keys Hurricane.

Hurricane planning became a part of the railroad construction process: Telegraph lines were installed so railroad supervisors could be kept aware of developing storms; well-built dormitories were constructed on the highest ground possible; and plans were put in place to sink critical equipment before the storms arrived. The category 3 storm in 1909 tore up forty miles of track in the Upper Keys, ripped free by the storm surge that covered the islands. As a result, the composition of the underlying roadbed was changed and a new mixture was laid to be sure the railroad would survive the hurricanes of the future.

What was originally called Flagler's Folly was, in the end, hailed as a miraculous achievement of engineering and determination. Henry Flagler died in 1913 at the age of eighty-three, just over a year after seeing his dream fulfilled, and never knowing, of course, that the dream would only last another twenty-two years. On Labor Day of 1935 the last train would roll into the Keys on a rescue mission. That train and the railroad through the Middle Keys would be wiped out that day by America's strongest hurricane and never rebuilt. The railroad was broke, financially strangled by the Great Depression. (See page 50.)

Second Galveston Hurricane
Landfall: Tuesday, August 17, 1915

The hurricane that proved that preparation works. After the cataclysmic storm of 1900, the city of Galveston undertook a hurricane preparation program that has never been duplicated. The city was reengineered, starting with a three-mile-long, seventeen-foot-high seawall along the Gulf. Even more dramatically, more than 2,100 buildings were raised, lifted up on stilts, so new land could be filled in underneath them. The test came in August of 1915. A hurricane—already at category 3 and intensifying—entered the Gulf of Mexico late on the fourteenth on a track similar to the 1900 storm's. Reaching category 4 strength on the sixteenth, the storm weakened a bit before sweeping across the southern end of Galveston Island early on the seventeenth. Winds were estimated at 120 mph; however, only 93 mph winds were measured at the Weather Bureau location in the city.

The sea wall held the storm surge back, but the waves went higher, crashing into the city and creating a torrent of water that washed away some of the new landfill. Houses crumbled and collapsed. There was significant damage to much of Galveston Island and the surrounding area, but only eleven people died in the city, largely because protection systems had been built. In addition, the storm was very well forecast. Hurricane warnings were issued at about 6:00 P.M. on the fifteenth and prompted many people to leave town. Still, the total death toll reached 275, with many of the fatalities occurring among people who stayed on boats.

Atlantic-Gulf Hurricane of 1919
Landfall: Dry Tortugas, FL, Wednesday,
September 10, 1919
Landfall: Corpus Christi, TX, Sunday,
September 14, 1919

The last major hurricane to hit the Lower Keys. On Monday, September 8, 1919, the pressure was falling and the winds were increasing at Miami. A growing and strengthening hurricane was moving by to the south. Hurricane warnings went up for the southern peninsula and the Florida Keys.

In Key West, the winds blew at tropical storm strength through the day Tuesday, gusting to 75 to 80 mph by early evening, blowing away the official anemometer. The worst weather came around midnight, when the center of the category 4 hurricane slid by fifteen to twenty miles to the south. The strong winds blew through the day on September 10, damaging virtually every building in the Lower Keys. The Weather Bureau reported, "The storm that passed over Key West . . . was, without a doubt, the most violent experienced since records at this station began." Over 13 inches of rain fell as well, adding to the misery, but there was no reported loss of life on land.

Moving slowly west-northwest, the eye of the still-strengthening hurricane passed directly over the Dry Tortugas, seventy miles west of Key West, near where the Spanish ship the *Valbanera* went down with all 488 aboard lost (the ship sank intact, but mysteriously the passengers were never found). Another ship measured a pressure of 927 mb about that time, only 5 mb higher than Hurricane Andrew, ranking this hurricane as the sixth strongest storm on record at the time of landfall in the United States.

The slow 5 mph forward speed prolonged the pounding in the Lower Keys and also forced Gulf Coast residents to wait for days to see what the storm would do next. By 1919, the Weather Bureau's warnings were getting wide distribution, which kept ships out of the danger zone. The problem was, ships' radio reports were the only way to locate the hurricane's center position. Through the weekend the entire Texas coast was on high alert, and Galveston went into full preparation mode, remembering well the devastating hurricanes of 1900 and 1915. On Saturday afternoon, September 13, forecasters had "lost" the storm. A monster hurricane was in the Gulf, but nobody knew which direction it was going.

On Sunday morning, September 14, 1919, the storm was "found" . . . just offshore of Corpus Christi. Evacuations of low-lying areas were under way—the weather was already deteriorating—but no one was prepared for what was to come.

The eye of the monster hurricane came ashore that afternoon about thirty miles south of the city, which meant that the east and northeast winds pushed the worst of the storm surge over the barrier islands and into Corpus Christi and the surrounding communities. Bodies of 121 people and 87 survivors were washed seven miles inland across Nueces Bay to the north of the city, where the water rose to the second floor of the downtown buildings. Debris piles were over fifteen feet high. Damage added up to about $20 million in 1919 dollars.

The official death toll from the great hurricane of 1919 is estimated to have been between 600 and 900. No accurate wind measurements were taken during the event—none of the Weather Bureau's instruments survived. Modern estimates are that the storm was at category 4 strength when it passed over the Dry Tortugas and a strong category 3 when it went ashore in Texas.

Tampa Bay Hurricane
Landfall: Tuesday, October 25, 1921

The last major hurricane to hit Tampa. Much like Hurricane Wilma in 2005, a tropical depression formed in the western Caribbean late in October of 1921. This storm also intensified quickly, reaching category 4 strength by the afternoon of October 23. But, unlike Wilma, the 1921 hurricane did not weaken over the Yucatán Peninsula, but instead threaded its way through the Yucatán Channel, maintaining its strength for the next day and a half. As with Wilma, the steering currents turned the storm toward Florida.

On the afternoon of October 25, 1921, the high-end category 3 storm came ashore over Tarpon Springs just north of Tampa. With landfall north of Tampa Bay, ahead of the storm the southwest winds estimated at 125 mph pushed a massive storm surge over the seawall and into the city. The storm tide was measured at 10.5 feet, the highest level since 1848. Homes and businesses near the water were destroyed throughout the Tampa Bay area, and in St. Petersburg the city's piers were ruined.

Over a hundred miles to the south, Sanibel Island was inundated, and in Fort Myers the water was higher than they had seen in over thirty years. The storm spread damage across the state, adding up to $3 million 1921 dollars. The center tracked over New Smyrna Beach on the east coast, but the winds by that time were down to category 1 strength.

Great Miami Hurricane
Landfall: Saturday, September 18, 1926

The most expensive U.S. hurricane if it were to happen today. Most of the 150,000 people in South Florida in 1926 had never been through a hurricane. A near miss by a storm in July convinced many people that they had little to worry about from tropical systems. Then, on September 14, the Weather Bureau became aware of a well-developed hurricane east of Puerto Rico. On the September 16, the storm—now at category 4 strength—decimated Grand Turk in the southern end of the Bahamas chain. On the morning of the September 17 no information was received from the Bahamas as was customary. Forecasters could only speculate where the hurricane was and where it was heading. At noon the Weather Bureau hoisted northeast storm warnings for Florida. That information reached the afternoon papers, but little attention was paid.

At 1:00 P.M. a report of north winds of approximately 50 mph finally came from Nassau. The barometer there was low and falling. A powerful hurricane was indeed in the vicinity. Through the afternoon and into the evening the pressure was *not* dropping rapidly at Miami and Key West, however. It was a difficult situation for a 1926 forecaster. At 10:00 P.M. the barometer in Miami began to fall. There is no indication that the Washington forecaster knew this, however. Simply based on his earlier analysis putting the hurricane near Miami at 8:00 A.M. the next morning—which turned out to be amazingly accurate—he issued hurricane warnings for South Florida at 11:00 P.M. on September 17. Warning flags were hoisted and calls were made to the telephone exchanges in the region, but most people had already gone to bed.

The center of the eye came ashore fourteen miles south of downtown Miami, but the monstrous eye of the even more monstrous storm still caused a thirty-five-minute lull in the city. Some people tried to cross the causeway to Miami Beach during the calm, only to be swept into Biscayne Bay by the storm surge that accompanied the second half of the storm. Nearly 95 percent of the buildings in what is now metropolitan Miami-Fort Lauderdale were damaged or destroyed. The ocean washed over Miami Beach and covered the island in sand, in some places several feet deep, according to the official report. The nearly 12-foot storm surge inundated the bay-front in downtown Miami and flooded hotels and homes and pushed boats of all sizes into the streets. Downtown Fort Lauderdale, twenty-five miles to the north, was also flooded by the massive storm. Maximum winds are estimated to have been 135 mph at landfall, although no anemometers survived to measure it.

San Filipe and Okeechobee Hurricane
Puerto Rico Landfall: Thursday, September 13, 1928
Florida Landfall: Monday, September 17, 1928

The first recorded category 5 hurricane. On September 10 word reached the Weather Bureau of a hurricane 1,250 miles east-southeast of Puerto Rico. This ship report is believed to be the first time information was ever received by radio about a hurricane far to the east in the Atlantic.

The storm steadily intensified, reaching the Leeward Islands as a category 4 hurricane on September 12. The eye passed directly over Guadeloupe, doing tremendous damage and killing 1,200 or more. Twenty-four hours later, now at category 5 strength, the storm took direct aim at Puerto Rico. It took nine hours for the massive hurricane to move diagonally across the island on a northwest heading.

Sustained winds in San Juan were measured at 159 mph (the first category 5 hurricane winds ever documented) over twenty-five miles north of the center, indicating that it's likely there were even stronger winds elsewhere in the circulation. (The San Juan wind reading may have been enhanced by the anemometer's location—research continues.) It was the most destructive hurricane ever to hit the island, but there were only 300 to 375 deaths, likely because of the advance warning by the Weather Bureau. See page 137 for more on the San Filipe Hurricane.

Lake Okeechobee disaster. The storm left Puerto Rico as a category 4 and maintained that strength until it reached the Florida coast three days later, 105 miles north of where the Great Miami Hurricane had come ashore almost exactly two years before. The eye came in north of Palm Beach, sparing the posh resort and nearby West Palm Beach from the worst of the wind. Still, both cities were smashed.

It's estimated that winds of 140 mph to 150 mph and a 10-foot storm surge accompanied the eye ashore. Forty miles inland, on the shores of Lake Okeechobee, a vibrant agricultural economy had attracted farmworkers to the rich land. As the storm moved west-northwest, the hurricane's winds pushed the water from the north end of the big lake against the south shore. It topped the 4-foot to 5-foot levee, flooding more than 400 square miles of farmland several feet deep. Untold thousands of farmworkers, mostly black, died in the flood. The exact number is unknown because many bodies were washed away. The official death toll is now set at over 2,500. It could easily have been 3,000 or more.

Florida Keys Labor Day Hurricane
Landfall: Monday, September 2, 1935

The strongest storm at landfall in United States history. A tropical storm drifted through the central Bahamas during the last days of August 1935, finally reaching category 1 hurricane status on Sunday morning, September 1, near the south end of Andros Island. As the small storm moved over the Gulf Stream, intensification became more rapid. The Weather Bureau was tracking the storm with "probably winds of hurricane force [in a] small area near [the] center." They thought it was moving due west and would likely stay south of the Keys. But the modern analysis is that by that evening it was already a category 2 hurricane and it had begun arcing to the west-northwest.

Labor Day afternoon, at 4:30 P.M. hurricane warnings were issued for the Florida Keys. The Weather Bureau now knew the storm was turning north. The winds were already approaching hurricane force in the Upper Keys and the barometer was dropping fast. Unbeknown to anyone, rapid intensification was continuing. By sunset, winds reached 160 mph, with gusts likely over 200. To the storm's west, the north wind pushed water from Florida Bay over the Keys, while to the east the ocean was piling ashore. The wall of water reached over thirty feet. In places the Keys were swept clean. Numerous clocks were recovered after the storm; all had stopped between 8:25 and 8:35 P.M. when the giant storm surge rushed in.

Down-and-out World War I veterans had been sent by the federal government to the Keys to build a highway to Key West. (One of the many projects the government sponsored during the Depression.) Most of them went to Miami for a Labor Day baseball game, but those who remained in the temporary camps on Upper and Lower Matecumbe Keys were in the direct path of the storm. A train came too late to rescue them. All of the cars except the locomotive were washed off the tracks. More than 250 veterans and 150 residents were killed by the storm. Estimates of the number of the dead run as high as 600. The total population of the Upper and Middle Keys at the time was less than 900 (not counting the veterans). Tens of thousands of people live there today.

Great New England Hurricane / Long Island Express
Landfall: Wednesday, September 21, 1938

From Florida to Long Island in one day. Early in the week of September 18 the existence of a major hurricane well east of the Bahamas was widely reported in the press. Residents in Florida—no doubt inspired by the cataclysmic storms of 1926, 1928, and 1935—were boarding up. The news on Wednesday morning, however, was that the large hurricane was swinging north and that the worst of it would stay offshore of the East Coast. Still, the Weather Bureau issues warnings from Virginia to New Jersey for the fringe effects of the storm. Now we know the monster hurricane had reached category 5 strength on the September 19 off the northeast Bahamas.

The weather had been cold and rainy for weeks on Long Island, New York. It was not a nice September at all. So, when more bad weather was forecast, there was no special concern. On Wednesday morning, September 21, the center of the hurricane was about seventy-five-miles off Cape

Hatteras, "moving rapidly north-northeast," according to the Weather Bureau advisory. The forecasters didn't realize, however, that it was *speeding* north at about 50 mph, so they were always behind the curve. The northern edge of the eyewall reached central Long Island, near Islip, at 1:50 P.M. The specific warning for that area went out at 3:00 P.M. By that time, the damage had been done.

The storm surge plus waves reached thirty feet or more on the south shore of Long Island. Coastal homes from Fire Island to the east were washed away. Most buildings on the eastern half of Long Island were severely damaged. But it was worse to the north. Like the Great 1815 Hurricane, Narragansett Bay, which opens to the south, was pushed into downtown Providence and covered the second floors of some buildings. The winds and coastal tides did widespread damage along the eastern edge of the storm track. Heavy rain, falling on the already saturated soil, caused flooding throughout the Northeast and New England that lasted many days. About 700 people died in the storm, 600 of them in New England. The modern analysis is that the storm likely became extratropical (see page 85) off southern New Jersey and came ashore with maximum sustained winds of about 100 mph. It had been extremely strong, however, with a large circulation—ingredients that produced a giant storm surge (like Hurricane Katrina in 2005). Providence recorded 87 mph sustained winds, and the peak wind was near hurricane force in both New York City and Boston. Blue Hill Observatory, near Boston, measured sustained winds of 121 mph with gusts to 186, but this reading was taken at an elevation of 635 feet. That intense circulation aloft was a legacy of a category 5 storm that had been east of Miami just the morning before.

1943 Surprise Hurricane
Landfall: Tuesday, July 27, 1943

Hurricane hunting is born on a bet. The system was first detected on July 25, 1943, in the Gulf south of Pensacola. America was in the middle of the Second World War, and although radar had been invented, it was not yet in use for storm tracking in the United States. Because of the fear of German U-boats in the

Gulf, all radio traffic from ships was silenced, including storm reports. Newspaper articles mentioned a tropical storm in the Gulf of Mexico, but did not indicate that the storm was a significant threat.

The hurricane made landfall midday on July 27 just north of Galveston, crossed Galveston Bay, and then tracked over Houston. Warnings were updated in the afternoon newspapers that day, but the storm was already bearing down, catching residents off guard. There are no official records of the storm, but newspaper reports indicate that the maximum sustained winds were about 86 mph at landfall.

Damage was significant, primarily from the wind, which reportedly gusted up to 132 mph at Houston's Metropolitan Airport (now Hobby Airport) and Baytown on Galveston Bay. The storm killed 19 people and caused $17 million (1943 dollars) in damage to the Houston area. After the loss of life in this storm, weather information has never been censored again.

The first hurricane flight(s). At Bryan, Texas, about eighty miles northwest of downtown Houston, U.S. Army Air Corps and British Royal Air Force pilots were in training, learning how to fly by instruments. At breakfast on the morning of July 27, the conversation reportedly turned to the airworthiness of the American AT-6 Texan Trainer, the aircraft being used in the training program. Major Joe Duckworth, the highly regarded creator of the instrument-flying program at the base, bet the Brits that he could fly through the approaching hurricane, which would prove the toughness of the AT-6.

Duckworth recruited Lieutenant Ralph O'Hair to be his navigator, and together they became the first men confirmed to have flown into a hurricane. While they were in the eye, they radioed the Houston weather office reporting the exact location of the center of the storm, something unknown in those days before storm-tracking radar. The storm was already ashore when Duckworth and O'Hair flew into it. Today, Hurricane Hunters stop at the coastline because of the severe turbulence that can develop over land, but nobody knew that in 1943.

Major Duckworth and Lieutenant O'Hair are regarded as the originators of hurricane hunting by aircraft. According to Ivan Tannehill's 1955 book, *The Hurricane Hunters*, several of Duckworth's instructors had flown into the same storm in B-25s, but were afraid to tell their boss. The practice of tracking storms by flying military aircraft through them soon became routine, dramatically increasing the knowledge we have of hurricanes and how they move.

Great Atlantic Hurricane
Landfall: Thursday, September 14, 1944
Havana-Florida Hurricane
Landfall: Thursday, October 19, 1944

Two monster hurricanes hit the United States. Hurricane season 1944 featured two hurricanes with huge circulations that did widespread damage. The first scheduled reconnaissance flight was sent into a developing storm north of Puerto Rico on September 9. By September 12 they were tracking a giant storm off the northeast Bahamas. The storm was so big and powerful that the Miami Hurricane Warning Office, the predecessor to the National Hurricane Center, named it the Great Atlantic Hurricane—the first "official" name issued by the Miami office.

The storm just missed Cape Hatteras, brushing by on Thursday morning, September 14, then sped north-northeast, coming ashore near Southampton on eastern Long Island late that same evening. The huge circulation spread destruction from North Carolina to Maine. Forty thousand or more homes were destroyed along the coast, but loss of life on land was relatively light because of the much improved warnings that were issued by the Weather Bureau after the "surprise" mega storm six years earlier. Ships patrolling the east coast were caught in the fierce winds and waves, however, and 390 sailors died when several sank, including a destroyer, the USS *Warrington*.

Exactly a month later another storm was forming in the western Caribbean, meandering around, growing in size, and dropping more than 31 inches of rain on the Cayman Islands. Finally the storm moved north, pushing a massive storm surge over Cuba's southern coastline. Wind gusts were measured at 163 mph in Havana.

The storm, now estimated to have been a category 3, took the typical October path to the north and northeast toward the west coast of Florida. The Weather Bureau report said, "dangerous winds extended fully 200 miles to the right or east of the center . . . thus affecting the entire peninsula." The track was similar to Hurricane Charley sixty years later, but this storm was many times bigger. Winds gusting to 100 mph at Orlando and Tampa and just under hurricane force in Miami did damage across the state. A causeway was washed out near Cape Canaveral, and a record tide of over 12 feet hit Jacksonville Beach. Florida's death toll was 18, but hundreds died in Cuba.

South Florida's Two Hurricanes and a Tropical Storm
Hurricane #1 Landfall: Wednesday, September 17, 1947
Hurricane #2 Landfall: Saturday, October 11, 1947

The highest "official" sustained wind measurement taken in the United States. The first hit in a one-two hurricane punch for South Florida in 1947 came from a monster category 4 storm that came ashore over "largely unpopulated" northern Broward and southern Palm Beach counties (as the local newspaper described the area at the time). Today's cities of Boca Raton, Deerfield Beach, Pompano Beach, Margate, Coconut Creek, and Coral Springs all lie directly in the track of the storm. About 500,000 people live in those cities today.

The disturbance was first detected over western Africa by the Pan American World Airways station in Dakar on September 2. Pan Am's observations were an important part of the weather monitoring system of the day. On September 10, a ship radioed that a well-developed tropical storm was approaching the Lesser Antilles. The storm swung north, however, toward the northern Bahamas and Florida, steadily intensifying into a category 5 monster just east of Great Abaco island at the top of the Bahamian chain. The huge hurricane moved slowly for a day before heading west-southwest across Abaco and into the southern Florida peninsula just north of Fort Lauderdale.

An official reading of 155 mph was taken at the Hillsboro Lighthouse in the small town of Hillsboro Beach, Florida (immediately north of Pompano Beach). This is the highest official *sustained* wind reading ever taken on land in the United States (a higher *gust* measurement was made during Hurricane Andrew and in numerous other storms). The anemometer, however, was located at the top of the lighthouse, about 140 feet above sea level. So adjusting for height using an 89 percent reduction factor (provided by the National Hurricane Center), the "official" sustained wind speed would be about 138 mph. See page 85 for more on "sustained" winds.

The monster storm spread hurricane-force winds over the entire southern half of the Florida peninsula. An 11-foot storm surge washed over the coast from Fort Lauderdale to Palm Beach. Even more than twenty-five miles south of the center's track in Dade County, the storm surge pushed into Miami Beach and A1A was washed out north of the city. The most severe damage was reported in Fort Lauderdale and Hollywood, but only because the area to the north, where the strongest winds were, was so lightly populated.

Major flooding was reported throughout Greater Miami. The summer of 1947 had been extremely wet, so the water table was unusually high before the hurricane hit.

Then *six days* later, a 50 mph tropical storm passed by the peninsula to the west and brought more rain to the Miami/Fort Lauderdale area.

Hurricane #2 arrives. Two and a half weeks after the brief skirmish with the tropical storm, and just three weeks after the category 4 hurricane blew through, another hurricane crossed South Florida, this time from the southwest. When this category 1 storm finally left town on October 12, most of metropolitan Miami/Fort Lauderdale was underwater.

The center of the storm came ashore at Cape Sable in extreme southwest Florida as a 75 mph hurricane and slowly intensified as it moved east-northeast toward the metropolitan area. The strongest winds reached the east coast after midnight. Amazingly, for the second time in less than four weeks the center of a hurricane went directly over the Hillsboro Lighthouse. This time, however, the maximum sustained winds were measured at about 80 mph, but there wasn't much left to damage.

The wind wasn't the story, however. Torrential rain fell through the night. In Hialeah, six inches fell in an hour and fifteen minutes. Five to 15 inches of rain fell across the southern part of the peninsula. U.S. 1, which was built along the coastal ridge (among the highest ground in South Florida), was underwater from Fort Lauderdale to Miami. Virtually every neighborhood was flooded, and the deep water extended across the state. It was said that you could take a canoe from Fort Lauderdale on the east coast to Naples on the west coast for weeks.

The massive flood prompted the construction of a sophisticated system of levees, dams, and canals run by the South Florida Water Management District. Over the past few years massive pumps were added to the system to lower the water table more quickly after heavy rain. The metropolitan South Florida area today is almost fully developed, however, so there is far less open ground where water can soak in than there was sixty years ago. Even with the high-tech system in place, there is every expectation that flooding will again be a major problem in a wet scenario like the one that developed in 1947.

Cloud seeding and the hard left turn. In July 1946, scientists working for General Electric discovered that clouds could be "seeded" to induce the water droplets to coalesce into snowflakes or raindrops. Theoretically, it was thought the same process could be used to weaken hurricanes by forcing them to "rain themselves out," cooling the inner core. A team of scientists from GE and the U.S. military finally got to test the theory on South Florida's 1947 hurricane #2. The storm was heading northeast, apparently on a course toward the hurricane graveyard of the North Atlantic, and therefore safe for experimentation. The program was called Project Cirrus.

On the morning of October 13, 1947, about 200 pounds of dry ice was spread through the storm, at that time located about 350 miles east of Jacksonville, Florida. The scientists on board the plane noticed that the appearance of the seeded clouds changed, but the overall effect on the storm could not be determined. Then came the big surprise.

Shortly after the seeding, the hurricane hung a hard left and headed directly for the Georgia and South Carolina coast. The 85 mph storm came ashore near Savannah on October 15, doing significant damage.

The Project Cirrus mission was classified so the public never knew about the seeding experiment, and General Electric escaped any lawsuits that might have come out it.

At the time, the lead GE scientist was sure the cloud seeding had done it. A later examination of the upper winds, however, showed that the storm's path was determined by the large-scale atmospheric pattern, as is almost always the case. Efforts to modify the strength of hurricanes would continue off and on until 1983 (see "Project Storm Fury" on page 101), but conclusive results would always be elusive.

**Palm Beach Hurricane
Landfall: Friday, August 26, 1949**

Last major hurricane in West Palm Beach. Two years after a powerful category 4 storm slammed southern Palm Beach County, another hurricane took a track across Florida frighteningly similar to the catastrophic Lake Okeechobee Hurricane of 1928. While this storm was fairly weak as it moved by north of Puerto Rico, it intensified into a 135 mph hurricane as it passed just north of Nassau early on August 26. Through that day it intensified to 150 mph, at the top end of the category 4 scale as it headed toward the Florida coast.

The eye passed directly over West Palm Beach, where the winds went calm at the airport—by modern reckoning about nine miles south of the track of the circulation center—for twenty-two minutes. Top sustained winds there were estimated at 120 mph. Up the coast some twenty miles at the Jupiter Lighthouse, top winds were measured at 153 mph before the anemometer failed, but that value was taken about 150 feet above sea level. The maximum landfall sustained wind speed is subject to speculation. Since the 120 mph estimate was south of the track, on the left side of the storm, it could be slightly low. Adjusting the Jupiter Lighthouse reading to a surface wind using the National Hurricane Center reduction factor yields 133 mph. When the data from this storm is reanalyzed by NOAA scientists, it will likely go in the record books as a high-end category 3 or low-end category 4 hurricane at landfall.

The most severe damage was reported to the north of where the center came ashore, in Jupiter and Stuart. The storm created a 12-foot storm surge on Lake Okeechobee, which challenged the Hoover Dike, which had been built around the southern end of the lake after the 1928 disaster, but the dike held. Major damage was done to the citrus crop to the north of the lake as the weakening storm arced in that direction.

In 1949, only 143,000 people lived in the three coastal counties hit by the hurricane. About 1.7 million live there today.

1949 footnote. Another major hurricane came ashore early on October 4, 1949, near Freeport, Texas. The track went directly over Houston, but by then the winds had decreased to 90 mph. It was the last hurricane to pass directly over the city until Hurricane Alicia in 1983.

Hurricane Carol
Landfall: Tuesday, August 31, 1954

The worst New England storm since 1938. Late on August 30 category 2 Hurricane Carol brushed Cape Hatteras before speeding toward the northeast. The storm's track was just east of the great hurricane of 1938, limiting somewhat the damage on Long Island. But Carol caused major devastation in eastern Connecticut, Rhode Island, and Massachusetts. Providence was especially hard hit, with severe flooding downtown reminiscent of the 1938 storm. Carol is considered the most damaging hurricane ever to hit Cape Cod.

Hurricane Edna
Landfall: Saturday, September 11, 1954

Major damage done in Maine and eastern Canada. Less than two weeks after Hurricane Carol, Hurricane Edna took a similar path, although just to the east. Cape Hatteras was hit again, but Edna's category 3 winds stayed offshore. The storm weakened slowly as it arced north-northeast. It just missed Long Island but hit the eastern end of Cape Cod and came ashore near Eastport,

Maine. Winds of about 75 mph and heavy rain did major damage. Edna is a reminder that even Maine is not immune to hurricanes.

Hurricane Hazel
Landfall: Sunday, October 15, 1954

The hurricane of record in North Carolina and Toronto. Just less than a month after Edna, Hazel, the strongest hurricane of the year, moved through the Caribbean. On October 9, the category 4 storm turned north toward the Windward Passage between Cuba and Haiti. For over a day the storm pounded the western end of the island near Port-au-Prince; its powerful winds, torrential rain, and storm surge spread death and destruction. Hazel weakened a bit during its interaction with the high terrain of Haiti and Cuba, but by October 14 it was back to category 4 strength and heading toward the East Coast of the United States.

On the morning of October 15 the eye came ashore at the South Carolina—North Carolina border, putting the strong right side of the storm and the eighteen-foot storm surge on the North Carolina coast. The category 3 winds and massive surge—which came at an astronomical high tide—obliterated "all traces of civilization" in the area south of Cape Fear, according to the Weather Bureau in Raleigh. The damage continued as the storm raced northward through North Carolina and Virginia and into the Northeast. A wind gust of 113 mph was recorded at the Battery, at the south end of Manhattan, the highest wind ever recorded in the New York City area. Late that night, Hazel crossed into Canada. Now an extratropical storm (see page 85), but still with winds at hurricane force and carrying tropical moisture, Hazel hit Toronto hard. Record rainfall of over 8 inches fell on already saturated ground. The raging Humber River swept away bridges, roads, and homes. All of the houses on Raymore Drive were washed into the river, along with 32 residents. In all, 81 people died in Canada, 95 in the United States, and 600 to 1,000 more in Haiti. Hazel was the most damaging hurricane ever to hit the United States up to that time.

An uncounted casualty of Hurricane Hazel was the chief hurricane forecaster at the Miami Weather Bureau (today he would have the title "director of the National Hurricane Center"), Grady Norton. In September he had been advised by his doctor to rest, but Norton could see that Hazel was going to be the worst hurricane of the year. After working a typical long day, Grady Norton suffered a stroke and died on October 9, just after Hazel reached category 4 strength for the first time.

Hurricane Connie, Diane, and Ione
Landfalls: August 12 and 17 and September 19, 1955

The East Coast hurricane onslaught continues. None of the storms was remarkable individually, although Connie and Diane caused record flooding in the Northeast and New England. The legacy of the 1955 storms, however, is that they came the year after the 1954 triple hit on the East Coast. Another hurricane, Barbara, had also taken a similar path in 1953. What did it mean? Were the weather patterns changing? The answer in the end was no. But, it is a reminder that Mother Nature does, on occasion, get stuck in a rut.

Hurricane Donna
Landfall: Saturday, September 10, 1960

The only storm to produce hurricane-force gusts in every east coast state from Florida to New Hampshire. For a record nine days from September 2 to the 11, Donna maintained its strength at category 3 or above as it moved across the tropical Atlantic and through the Bahamas toward Florida. In the early morning hours of September 10, the category 4 storm came ashore over Marathon (actually Duck Key, just to the north on the same island) in the Middle Florida Keys. The 135-mph sustained winds and 13-foot storm surge destroyed all but the most substantial buildings from Marathon to

Tavernier, forty-five-miles to the northeast. The monstrous circulation curved north across Florida from near Naples to Daytona Beach. The entire peninsula was affected, including the ecosystem of the Everglades. President Dwight Eisenhower declared all of Florida south of Orlando a disaster area.

Donna continued north, making a third landfall in eastern North Carolina on September 12. It quickly exited into the Atlantic again and raced toward New England. After a fourth landfall on the eastern end of Long Island, the huge system sped by just to the west of Boston late that evening. In the Mid-Atlantic and Northeast Donna produced moderate but widespread damage from its category 2 winds and very heavy rain.

Hurricane Carla
Landfall: Monday, September 11, 1961

The hurricane that made Dan Rather famous.
Hurricane tracking turned a big corner in 1961 when the Tiros III satellite was launched in July. While hurricanes had been seen from space since the previous year, a storm was discovered by satellite for the first time on September 11, 1961. As that new storm, named Esther, was developing in the distant tropical Atlantic, giant Hurricane Carla was coming ashore in Texas.

The water was already rising on the Texas coast on Friday, September 8, with Carla still near the Yucatán Peninsula, 600 miles offshore. The

1961 Hurricanes Betsy, Carla, and Debbie from NASA's Tiros III Satellite.

monster storm was already at category 3 strength and getting stronger. The following day, the entire Texas coastline was put under a hurricane warning, prompting the largest evacuation ever undertaken in the United States to that time. Some 350,000 people fled inland as the height and the power of the waves battering the coast increased over the next two days.

The slow-moving storm grew in size and strength, reaching category 5 the evening of September 10 when the center was 150 miles east-southeast of Corpus Christi. And it continued to strengthen, its circulation enveloping the entire Gulf of Mexico. Maximum sustained winds were estimated at 175 mph overnight as Carla's eye relentlessly crept toward the same part of Texas that had been decimated in 1875 and 1886. (See the Indianola Hurricane on page 38.)

Dan Rather was working for KHOU-TV in Houston in 1961, reporting from the Weather Bureau office in Galveston. As the storm crept closer to the coast, Rather interviewed the meteorologists and showed the live radar on the air, describing the immense size and power of the storm. His broadcasts were widely acknowledged as significantly aiding the evacuation. CBS News took note of his work, and later that year he was working for the network.

Carla weakened slightly, creeping ashore as a 145 to 150 mph category 4 hurricane near Port O'Connor, Texas, about fifty miles up the coast from Corpus Christi. If Indianola had been rebuilt after the 1886 disaster, it would no doubt have been wiped out again.

1961 footnote. Hurricane Debbie took an unusual track in the eastern Atlantic, hitting Ireland as a category 1 hurricane. Debbie is the only hurricane ever to hit Europe, according to the official records going back to 1851. Vince hit Spain as a tropical depression in 2005 (see page 78), the only other system to survive the trip to the European coast as a tropical cyclone.

Hurricane Cleo
Landfall: Thursday, August 27, 1964

The last hurricane eye to pass over downtown Miami. Hurricane Cleo reached category 4 strength in the eastern Caribbean on August 22, 1964, after blasting the southern part of the French island of Guadeloupe. Hundreds of homes were demolished and 14 people died in the capital of Basse-Terre. On August 23 with Cleo continuing west-northwest through the northern Caribbean, a Navy Super Constellation Hurricane Hunter plane was badly damaged (it had to be scrapped) and seven crewmen were hurt as they tried to measure the intensity of the strengthening hurricane.

After mauling Les Cayes on the southwestern peninsula in Haiti, the small but vicious storm moved on toward Cuba, making a northward bend as it approached the southern coast. (The turn was unexpected and never explained based on the meteorological data available at the time.) The mountains of Haiti and eastern Cuba took their toll on Cleo's circulation, so the system that headed north toward Florida on August 26 was barely a hurricane.

Cleo was constantly monitored by aircraft, and eventually by radar, through the afternoon and evening of its trip over the Florida Straits. When the plane finally left the storm at about 11:00 P.M. on Wednesday,

August 26, winds on the western side of the circulation were measured at less than hurricane strength and the pressure was "stable" at 984 mb.

In the early-morning hours of August 27, the eye of a much stronger Hurricane Cleo came ashore on Key Biscayne and moved over downtown Miami. The pressure had dropped dramatically, and the winds had increased to 100 mph or more in all quadrants of the storm in those three hours before landfall. Hurricane warnings had been issued for the entire southeast coast, but most of South Florida was not ready for a direct hit from a strong category 2 hurricane.

Former NHC director Neil Frank remembers that then director Gordon Dunn announced that the eye was expected to stay twenty-two miles offshore with 45 to 50 mph winds affecting the coast. (The 8:00 P.M. advisory said, "The anticipated path will carry the heaviest weather and strongest winds a short distance offshore from Miami a little after midnight.") In a lesson that resonates today, overly precise hurricane forecasts can do more harm than good.

Cleo was the first hurricane to be tracked by two satellites (Tiros and Nimbus I) and the new high-powered radar system recently deployed by the Weather Bureau. Perhaps the new technology encouraged forecasters to be more specific than they would previously have been. In any case, Florida senator George Smathers called for an investigation into the cause of the "misinformation" that misled area residents to see if additional facilities or equipment should be added to the Miami weather office.

After coming ashore over Miami, Cleo turned slightly north, spreading hurricane-force winds up the east coast to the Melbourne area in central Florida. Damage added up to about $125 million in 1964 dollars. The only significant injury was to a guest at the posh Fontainebleau Hotel on Miami Beach—a broken arm from a flying door. But damage was widespread. Plate-glass windows crashed into lobbies along the beach, planes were overturned, a TV tower toppled, power and phone service was out for days, and the *Fort Lauderdale Daily News* couldn't print for the only time in its history.

 1964 footnote #1: Hurricane Dora. Cleo was just the first of three storms to hit Florida in 1964. Two weeks later, category 2 Hurricane Dora came ashore over St. Augustine, the only storm on record to hit the northeast Florida coast moving east to west. The long duration of onshore winds pushed high water ashore from Daytona Beach north into Georgia, washing out roads and smashing coastal homes and businesses. Winds were estimated at 125 mph in St. Augustine and near 100 mph along the coast north of the storm center through the afternoon and evening of September 10, 1964. Damage estimates ranged from $200 million to $240 million, including agricultural damage in both Florida and Georgia. The lasting legacy of Dora is the reminder that Florida's northeast coast is not hurricane-proof after all.

 1964 footnote #2: Hurricane Isbell. Then, a month later, on October 14, 1964, Hurricane Isbell came ashore on Florida's southwest coast and moved quickly across the southern part of the peninsula, exiting just north of Palm Beach. Winds were measured at 90 mph at both Everglades City, near the landfall point, and Indiantown, twenty-five miles northwest of the exit point on the east coast. The storm weakened as it headed north over the Atlantic, but flooding rain affected eastern North and South Carolina and eastern

Virginia. Most of the $6 million in damage from Isbell was to agriculture in Florida and North Carolina.

1964 footnote #3. Hurricane Hilda. Between Dora and Isbell, category 3 Hurricane Hilda moved out of the Caribbean and made landfall on the southern coast of Louisiana ninety miles west-southwest of New Orleans on October 3, 1964. An almost complete evacuation of the coastal area prevented a large loss of life; however, tragedy struck the small town of Erath, Louisiana, just northwest of Hilda's landfall point. Late in the afternoon, as Hilda was approaching, the town's water tower collapsed, crushing City Hall, where Civil Defense was being coordinated. Eight people died in the ruins of the one-story brick building.

Hurricane Betsy
Florida Landfall: Wednesday, September 8, 1965
Louisiana Landfall: Thursday, September 9, 1965

The first billion-dollar hurricane. After fits and starts, including the first of two loops, on September 1, 1965, Betsy was at hurricane strength and strengthening well northwest of Puerto Rico. Concern was increasing from South Florida to the Carolinas. By Friday morning, September 3, Betsy was 450 miles due east of Miami with top winds around 125 mph, but the track was now clearly taking the storm to the north toward the Carolinas. The chief forecaster in Miami, Gordon Dunn, announced on television that afternoon that the threat to southeast Florida was over. North Carolina was alerted that a hurricane watch might be necessary that night.

The following day, with top winds now at 140 mph, Betsy slowed its forward progress and had, by evening, begun its second loop. By early the next morning, Betsy was as far north as it was going to go, 320 miles east of Cape Canaveral (at the time called Cape Kennedy). Through the day and night it moved—unbelievably to shocked Bahamians and South Floridians—*south*, arriving just offshore of Great Abaco in the northeast Bahamas as a 115 mph hurricane Labor Day morning, September 6. A hurricane warning was issued for South Florida at 11:00 A.M.

Betsy crept along, pounding the Abacos for twenty hours, with winds reportedly between 135 and 147 mph much of the afternoon. On Tuesday the seventh, incredibly Betsy stalled again, crawling west just fifteen miles north of Nassau. The eye had expanded to forty miles in diameter, so in the Bahamian capital the winds went "dead calm" for hours in the afternoon. The island was slammed by the eyewall in the morning and then again at night. Residents

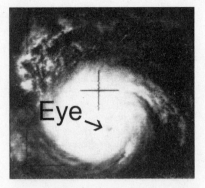

Hurricane Betsy viewed by NASA's Tiros X satellite on September 3, 1965, when it was about two hundred miles east of the northern Bahamas and forecast to head toward North Carolina.

were holed up in their homes or shelters for two nights and part of three days. No official wind measurements were taken—the instruments failed—but modern estimates are that Betsy's top winds were 115 mph as it crept by Nassau. During the night of September 7 and into the early-morning hours of the following day, the category 3 storm was getting stronger and finally making a move toward Florida. A "hurricane emergency warning," as they called it then, was in effect for the southern peninsula.

Throughout the slow trek through the Bahamas, relentless northeast winds had pushed water toward the Florida coast. Now Betsy was going to bring a storm surge on top of the already high tide. By late on Tuesday, September 7, wind had reached 81 mph at Miami Beach and 83 mph in Fort Lauderdale, even though Betsy's eye was heading for the Upper Keys, some fifty miles south of Miami. About 3:45 A.M. on September 8, 1965, Betsy made its first U.S. landfall over Key Largo with winds of approximately 125 mph; the Keys were shredded; the Overseas Highway was once again broken in six places, as it had been just five years earlier in Hurricane Donna. Near Miami, Key Biscayne flooded, killing hundreds of animals at the city's zoo, which was located on the island. Hotels along the coast were pounded—as they had been by Cleo the year before—with waves washing

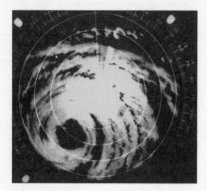

Hurricane Betsy viewed from Miami
Weather Bureau radar.

into some lobbies. Coastal areas were flooded as far north as Melbourne, over two hundred miles up the coast. And there was much more to come.

Betsy heads into the Gulf. On Thursday, September 9, Betsy strengthened and picked up forward speed. Early that morning hurricane warnings were issued from southeastern Louisiana to Galveston, Texas. The thinking was that the storm would go well west of New Orleans. By late that afternoon, however, the forecast had changed and New Orleans was told to get ready. At 7:00 CT that evening, the massive, fast-moving hurricane was just offshore of the southeastern tip of Louisiana, about 125 miles southeast of the city. Top winds were 150 to 155 mph.

Through the night the wind roared—sustained winds were estimated at 125 mph at the New Orleans Weather Bureau just before midnight—and the storm surge poured into the bayous, bays, and canals in Louisiana and Mississippi. Betsy's track took the center west of New Orleans, meaning the huge circulation pushed Gulf water up the Mississippi River as well as into the east side of the city and Lake Pontchartrain. Betsy's eyewall winds battered the metropolitan area until the early-morning hours.

In a preview to what we would see forty years later, levees along the Mississippi River–Gulf Outlet canal and the Industrial Canal were overtopped and breached, pouring water into the Ninth Ward and nearby neighborhoods. Except that the pictures are in black and white, and the swamped cars are from the 1950s and '60s, the scenes of houses flooded to their roofs look identical to those created by Hurricane Katrina. The difference in Betsy was that there were no winds to drive Lake Pontchartrain's water into the city from the north, so the flooding was not as widespread.

Hurricane Betsy was then the most expensive hurricane, by far, with damages adding up to about 1.5 billion 1965 dollars—the first billion-dollar disaster in U.S. history. Encouraged by the political persuasiveness of Louisiana senator Russell Long—the tape of the phone call still exists—President Lyndon Johnson was on the ground touring the devastated areas the night after the hurricane hit. "This is your president, I'm here to help you," Johnson shouted as he entered a dark and crowded shelter in the Ninth Ward with only a couple of flashlights to lead the way.

Johnson promised that federal red tape would be cut and New Orleans would not be forgotten. The Army Corps of Engineers undertook a massive program to raise and strengthen the levee system protecting the city, but, as we found out in 2005, mistakes were made that contributed to an even greater disaster when the next big hurricane hit.

Hurricane Camille
Landfall: Sunday, August 17, 1969

The strongest hurricane to hit the United States in modern meteorological times. Camille was first detected by satellite near Grand Cayman Island on August 14 and was immediately named a 60 mph tropical storm. Two days later the storm had crossed the western end of Cuba and was strengthening dramatically in the Gulf of Mexico. The evening of August 16, Hurricane Hunters measured winds of 160 mph. Before that day, a category 5 hurricane had never been seen in the Gulf of Mexico. A massive evacuation effort was undertaken on the central Gulf Coast. Camille moved fairly slowly north-northwest, the center arriving near Bay St. Louis, Mississippi, late on September 17. Incredibly, the storm strengthened even more. Sustained winds are estimated to have been 190 mph at landfall. It is extremely unusual for a storm to maintain itself at maximum strength for so long. The destruction along the Mississippi coast was total. From the Gulf, as far as you could see inland, there was nothing left. There was complete destruction even a half mile from the water. Almost everyone who did not evacuate the waterfront was killed. The total death toll is estimated to have been 255.

Camille's storm surge of 23 feet, wind of 190 mph, and lowest central pressure of 905 mb (909 mb at landfall) all set records. It is still one of the strongest hurricanes ever measured, although Hurricane Katrina produced a higher storm surge in virtually the same spot in 2005. Camille was a lot smaller than Katrina in diameter, but it was much, much stronger. The 1935 Keys storm, Hurricane Camille, and Hurricane Andrew are in a class of their own: small but extraordinarily powerful storms that came ashore at full strength.

Hurricane Hugo
Landfall: Sunday, September 22, 1989

First "modern" United States hurricane disaster.
Computer power had increased enough by 1989 that hurricane forecasting was moving into the modern era. Computer models performed reasonably well on Hugo, and the National Hurricane Center's forecast errors were significantly less than in the previous ten years. Communications systems were mature, and the evacuation of the South Carolina coastal areas went smoothly. Still, Hugo's damage had triple the price tag of the previously most expensive U.S. hurricane, Frederic, in 1979.

Category 4 Hurricane Hugo took a path through the northeastern Caribbean and did damage on a scale not seen in modern times. On the seventeenth, Hugo's eye went directly over Guadeloupe and St. Croix, in the Virgin Islands, and the northeast corner of Puerto Rico causing tremendous destruction. Then the storm swung northwest toward the East Coast. Hurricane warnings were issued from North Florida to North Carolina on the morning of the September 21. The center of the 140 mph storm came ashore late that night, officially at midnight, over Sullivan's Island, just north of Charleston. That slight deviation to the north spared the historic city; there the top recorded winds were only about 90 mph. Coastal communities for a hundred miles north of Charleston sustained tremendous damage. The storm surge was as high as 20 feet. The huge circulation maintained significant strength well inland. Charlotte, North Carolina, 170 miles from Charleston, recorded a wind gust of 99 mph.

Hurricane Hunters' Big Scare

On September 15, satellite estimates were that Hugo was a category 2 storm well east of the Lesser Antilles. A NOAA reconnaissance flight reached the storm late in the day and elected to fly into the eye at 1,500 feet, something done only in relatively weak hurricanes. Flights at lower altitudes (especially in those days) yield better data, but leave little room for error. This sortie turned out to be a big mistake. The P-3 aircraft was tossed and twisted beyond its design specifications. One engine caught fire. Hugo had intensified to category 5 strength. The crew made it into the eye and calm winds, got control of the aircraft, and eventually limped back to base. These days Hurricane Hunters normally stay at 10,000 feet.

Hurricane Mitch
Landfall: Wednesday, October 29, 1998

The first mega hurricane of the current high-activity period. In the last few years we have gotten used to seeing large hurricanes strengthening rapidly and Hurricane Hunters reporting stunning low-pressure numbers. (This is *not* a good thing!) Katrina, Rita, Wilma, and Ivan all fall into this category. Mitch was the first storm to exhibit this breathtaking trait since Atlantic hurricane activity increased dramatically in 1995. Beginning on October 24, Mitch strengthened at a spectacular rate, reaching category 5 strength 110 miles off the northeast Honduras coast on October 26. Hurricane Hunters measured 180 mph winds and a 905 mb central pressure. This was the lowest pressure ever seen in an October hurricane, until Wilma in 2005.

Mitch drifted westward then eventually southwestward for three days, in spite of forecasts that showed the storm turning north. The circulation's interaction with the Honduran mountains and other factors weakened the system until it finally moved ashore on the northern Honduras coast. Rainfall totals of three feet or more fell over the mountainous terrain as the weakening storm moved slowly over Honduras and Nicaragua. Massive mudslides swept down mountainsides throughout the region. Houses with their residents inside slid into the raging Choluteca River in Tegucigalpa, the capital of Honduras, when hillsides gave way. In all, 9,000 to 10,000 people died in Honduras and Nicaragua. Many bodies were never found, no doubt because they were buried in the mud. Mitch ranks as the third- or fourth-deadliest hurricane on record. Ironically, Hurricane Fifi caused approximately the same number of deaths from torrential rains over Honduras in 1974.

Other Notable Hurricanes

The list above is far from complete. I can hear students of hurricane history rattling off a list of major hurricanes that, at the time, were local catastrophes. The great Gulf Coast hurricane of 1906 that demolished Mobile and Pensacola (the worst in 170 years), and the Great New Orleans Hurricane of 1915, are examples. (Interestingly, both occurred in years with other storms listed above.) The storms in this chapter, however, are chosen to be representative, not just individually, but also in the trends they indicate. Hurricanes occur in streaks. Having one bad storm doesn't mean you're not going to have another next week, next month, or next year. Hurricanes occur in many places where people feel immune, such as New York City. And hurricanes can change things forever in a coastal area. Galveston, for example, never again was the major port it had been before the hurricanes of 1900 and 1915. Houston developed instead.

I've included more than an average number of Miami-area storms because an inordinate number of hurricanes hit there in the last one hundred or so years. Many South Floridians think that the storms of 2005 were a fluke, but reading through the accounts of the handful of major hurricanes listed above that have hit the southern Florida peninsula might convince them otherwise. And there were more storms that didn't make the list.

The overriding message to take from the accounts above has to be: If you live in the hurricane zone, you live in a place with a hurricane history. And, when history repeats itself, the damage will be dramatically worse due to the increase in people and wealth at America's coastlines.

Oddball Tropical Systems

Southern California Hurricane
Saturday, October 2, 1858

The only hurricane known to have had a direct impact in Southern California. Reports from California newspapers of the time tell of a "terrific and violent hurricane" hitting San Diego and the Southern California coast. Meteorological records imply that the eye never came ashore, but "extensive damage was done in the city" (which was really a small town with fewer than 1,000 residents). Some homes were reported to have collapsed.

A strong El Niño was believed to have been under way in 1858. Uncommonly warm nighttime temperatures had been reported that fall, perhaps because the ocean-water temperature was well above normal. The effects of what is believed to have been a category 1 hurricane (as it passed offshore of San Diego) were felt throughout Southern California. At San Pedro, boats were tossed ashore and heavy rain fell throughout the region. Parts of the Los Angeles area also reported gusty winds.

Long Beach Tropical Storm Landfall
Monday, September 25, 1939

The only known tropical storm to come ashore on the West Coast of the United States. Normally, when Pacific hurricanes take a northward track, they die out because the ocean water off Southern California is quite cool. During the El Niño year of 1939, the water was warm enough that a storm was able to maintain some of its strength as it moved north.

On September 25, a 50 mph tropical storm came ashore at Long Beach, the only tropical storm or hurricane ever to make landfall in California. Rainfall ranged from about 5 inches in the Los Angeles basin to a foot in the surrounding mountains.

Alice2 Becomes a Hurricane
Friday, December 31, 1954

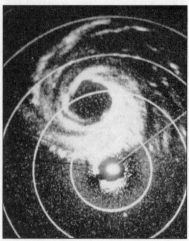

First there was Alice, then there was Alice2. The only January hurricane on record received the first 1955 name, Alice. Subsequent information showed that the storm had developed in the northeast Caribbean on December 30, 1954, however, and reached hurricane strength the next day. So, technically, the storm belonged on the 1954 list.

At that time, the naming system was in flux. The same list was being used each year, beginning with Alice. When this Alice was retroactively moved into 1954, which already had an Alice, this storm became Alice2. See page 136 "Hurricane Names."

Radar view of Hurricane Alice taken January 1, 1955. U.S. Navy photograph.

Alice2 had 80 mph winds when it passed through the northern Leeward Islands on January 3, 1955.

The Hattie, Simone, and Inga Trifecta
Tuesday, October 31, 1961

Three storms in one? With winds at or near category 5 strength, Hurricane Hattie came ashore just south of Belize City, blasting the capital of what was then called British Honduras (south of Cozumel, Mexico, on the Yucatán Peninsula). The city was 75 percent destroyed and about 275 people died. In 1970 the seat of government was moved to higher ground fifty miles farther inland.

Hattie quickly died out over Central America the next day. At the same time Tropical Storm Simone was developing 175 miles to the southwest on the other side of the Sierra Madre mountains on the Guatemala coast. Many texts indicate that Simone came from the remnants of Hattie, but that does not appear to be the case.

Simone moved north across Mexico and died out near the Gulf Coast on November 3. A day and half later Tropical Storm Inga formed 150 miles to the north, perhaps from Simone's remnants.

It's fun to think that all of these storms were really the same system. As I noted, it's likely not true. A full reanalysis of all of the available data will be undertaken by NOAA in the next couple of years. We'll await that work for the final word.

Super Typhoon Tip on October 13, 1979, when it was about 625 miles WNW of Guam. A day later, the central pressure was the lowest ever measured at sea level. Highest sustained winds peaked at 190 mph.

Super Typhoon Tip, the Granddaddy of Them All
October 1979

Tip's circulation would have stretched from Miami to Maine, almost 1,400 miles. The storm that would become one of the most incredible tropical cyclones ever seen had a difficult childhood. In early October of 1979 an elongated area of low pressure covered the Pacific Ocean from the Marshall Islands to the Philippines (technically it's called a monsoon trough). Three tropical cyclones formed out of this trough (not an uncommon occurrence), and two of those, Roger and Tip, were reasonably close to each other—close enough so that Roger's relatively weak circulation kept Tip from getting started. But once Roger moved out of the way on October 8, the weather pattern improved and Tip started to intensify.

After Tip passed just south of Guam as a tropical storm, the pattern became favorable for strengthening, and a typhoon was born (meaning winds reached hurricane strength). On October 11, Tip's highest winds reached 150 mph, the pressure was down to 898 mb, and the circulation extended 1,380 miles across, by far the largest ever measured. But Tip wasn't done.

Late in the evening of October 12, 1979, a reconnaissance aircraft measured Tip's central pressure at 870 mb, the lowest ever observed. This was 6 mb lower than the previous record set by Super Typhoon June in November 1975. Top winds were estimated at 190 mph.

Tip slowly weakened over the next week and came ashore on the Japanese island of Honshu about seventy miles south of Osaka. Winds had decreased to about 80 mph.

Have there been storms stronger than Tip? Super Typhoon Tip was closely monitored by aerial reconnaissance, which allowed direct measurement of the storm's winds and central pressure. Routine reconnaissance flights ended in August 1987, however. These days, the strength of tropical cyclones in the western Pacific is estimated by satellite.

A comparative analysis of the satellite pictures of Tip and subsequent superintense typhoons shows that two storms, Super Typhoon Gay in November 1992 and Super Typhoon Angela in November 1995, *may* have been even stronger. Top winds are estimated to have been 195 to 200 mph in these storms, but no direct measurements were taken.

Super Typhoon Angela hit the Philippines on November 2, 1995, doing considerable damage. By that time, however, the winds were estimated to have dropped to 160 mph. Super Typhoon Gay weakened dramatically before hitting Guam as a 100 mph storm.

Hurricane (Cyclone) Catarina Makes Landfall in Brazil
Sunday, March 28, 2004

If it quacks like a hurricane, it's a hurricane. Since the satellite era began in the mid-1960s, no hurricane had been seen in the South Atlantic until Catarina. Normally, the upper-atmosphere winds are too strong and often the water temperatures are too cool to support tropical cyclones. It's extremely rare for the atmospheric conditions to come together to allow tropical development.

On March 20, 2004, an extratropical low-pressure system moved off the southern Brazil coast and continued southeast for the next two days. On March 22 it stalled over moderately warm water, about 77°F. The atmospheric pattern became favorable, and the storm quickly developed into a hurricane and drifted back toward land. On Sunday, March 28, 2004, the first South Atlantic hurricane on record came ashore in the Brazilian state of Santa Catarina.

Hurricane Catarina as seen from the International Space Station. Courtesy NASA.

The strongest winds measured on land were 63 mph, but it's likely that sustained winds of over 75 mph came in with the storm. At least three people died and there was widespread destruction to homes and crops.

At the time, the Brazilian Meteorological Service would not classify it as a tropical cyclone, but the U.S. National Hurricane Center has confirmed that the system had the technical characteristics of a hurricane.

There is no official naming system for South Atlantic tropical cyclones. This one got its name because it made landfall in the state of Santa Catarina. Other hurricane-like storms that form south of the equator, mostly near Australia, are called cyclones, so this storm is sometimes called Cyclone Catarina.

Other South Atlantic Tropical Cyclones

A strong tropical depression/weak tropical storm was seen by satellite drifting away from the coast of Africa in April 1991. And another similar system appeared in January of 2004, two months before Catarina, off the central Brazil coast. It made landfall as a weak system. Then in February 2006 a system showed up on a satellite that appeared to be trying to organize into a tropical cyclone, but it did not. The odds, of course, favor the existence of other tropical cyclones in the South Atlantic in the years before satellites. They clearly don't happen often, however, and would likely have been dismissed as just a garden-variety bad storm if one had managed to come ashore.

Vince Makes Landfall in Spain
Tuesday, October 11, 2005

Wrong way, Vince! "If it looks like a hurricane, it probably is, despite its environment and unusual location." So said the National Hurricane Center on October 9, 2005, when it upgraded Vince

Hurricane Vince 615 miles southwest of the Iberian Peninsula, October 8, 2005. Courtesy NASA.

to a hurricane, the farthest east of any hurricane on record. The ocean water temperature was only about 74°F, and the upper-atmosphere pattern seemed only marginally favorable. But, the satellite presentation was impressive.

Vince weakened as it headed toward Spain and Portugal, making landfall near Huelva, Spain, on October 11 as a tropical depression. Still, Vince is the only tropical cyclone ever to come ashore in that part of the world.

Delta Hits the Canary Islands
Monday, November 28, 2005

Another wrong-way storm. At the end of the excruciating hurricane season in 2005, there was another odd storm in the eastern Atlantic. Like Hurricane Vince, Delta formed from an extratropical low-pressure system. As it moved over warm water, it gained tropical characteristics and was designated Tropical Storm Delta on November 23.

The storm drifted slowly for a few days, then interacted with an upper-level system and was moving to the east at 30 mph by Monday November 28. The storm had lost its tropical characteristics by that time, but still had winds of at least 65 mph when it hit the Canary Islands that evening. There was widespread damage on the islands of La Palma and Tenerife. Thousands of people had to be housed in emergency shelters. The remnants of Delta finally came ashore in Morocco.

Tropical Storm Delta near hurricane strength, November 24, 2005. Courtesy NASA.

Beware of Hurricane Records

We hear the words *record setting* a lot these days. The long list of records set during the 2005 hurricane season appears on page 15. There is no doubt that 2005 was an unusually busy season with exceptionally strong storms. However, we don't know how "unusual" and "exceptional" it was in reality. While reasonably reliable hurricane records go back to 1851, let's not kid ourselves. Before about 1950, airplanes were not regularly sent to investigate hurricanes. Before the early to mid 1970s, there were no regular satellite pictures. Before 1997, the instruments that were dropped into hurricanes were not as precise as those in use today. And we've only had wind data from the ocean surface measured by a high-tech radar system for the past few years. So, how many, how long-lived,

---

Sources: The details of the effects of past historic hurricanes are often quite sketchy and sometimes sensationalized. A landmark book on the subject, *Early American Hurricanes, 1492–1870*, by David Ludlum, was the source of many of the quotes I cited. Significant research reanalyzing hurricanes of the past—using historical records and a modern understanding of hurricane science—has been done in the past few years by Chris Landsea and a team at NOAA, complementing earlier work by Ed Rappaport and José Fernández-Partagás, who looked at storms going back to 1492. I've relied on their work, original government source material, and contemporaneous newspaper reports whenever possible.

and how strong hurricanes were in a given season was not precisely measured until recently.

Another issue complicating hurricane records is caused by Mother Nature's insistence on making things complicated. Most storms fall into one of two categories: tropical (powered by heat from the ocean) or extratropical (powered by energy from contrasting warm and cold air masses). But there are storms that have some of the attributes of both types. These days, these systems get named just like a tropical system, except they are called Subtropical Storm Whatever. (See page 142.) Previously, many or most of these systems might not have been counted and certainly wouldn't have been named. Thus it "seems like" there are more storms now, and this makes it even more difficult to accurately compare hurricane seasons now with those of the past.

Record-Setting Hurricanes

Most Intense Atlantic Hurricanes
(as determined by central pressure)

Storm Name	Year	Lowest Pressure (mb)
1. Wilma	2005	882
2. Gilbert	1988	888
3. Labor Day Keys	1935	892
4. Rita	2005	897
5. Allen	1980	899
6. Katrina	2005	902
7. Camille	1969	905
8. Mitch	1998	905
9. Ivan	2004	910
10. Janet	1955	914

Note: Use this table with caution. Notice that "intensity" is measured here by the storm's central pressure, not its highest winds. Also, the numbers given are the lowest pressures for which a credible measurement exists. For storms that occurred before about 1950, it was rare to get an accurate pressure reading until landfall. The 1935 Keys hurricane, for example, might have had lower pressure before it hit. We don't know. Thus the subtitle of the table.

Most Intense U.S. Hurricanes at Landfall

Storm Name	Landfall Location	Year	Landfall Pressure (mb)
1. Labor Day	Middle Keys	1935	892
2. Camille	Bay St. Louis, MS	1969	909
3. Katrina	Buras, LA	2005	920
4. Andrew	Homestead, FL	1992	922
5. Indianola	South Texas	1886	925*
6. Lower Keys	Key West, FL	1919	927†
7. Okeechobee	Jupiter, FL	1928	929
8. Donna	Marathon, FL	1960	930
9. New Orleans	Grand Isle, LA	1915	931
9. Carla	Port O'Connor, TX	1961	931

*Pressure estimated.

†Measurement was taken on a ship near the Dry Tortugas.

Most Expensive U.S. Hurricanes Based on Damage Caused When They Happened

Storm Name	Year	In Billions of 2005 Dollars (rounded to nearest billion)
1. Hurricane Katrina	2005	81+
2. Hurricane Andrew	1992	36
3. Hurricane Wilma*	2005	21
4. Hurricane Charley	2004	15
5. Hurricane Ivan	2004	15
6. Hurricane Hugo	1989	13
7. Hurricane Agnes**	1972	12
8. Hurricane Betsy†	1965	11
9. Hurricane Rita	2005	10
10. Hurricane Frances	2004	9

*Wilma's total is based on doubling the insured losses of $10.3 billion, a standard way of estimating the overall cost of modern hurricanes.

**The damage was not done when Agnes was a hurricane. The remnants caused massive flooding in the Northeast.

†Betsy was the first hurricane to cause more than a billion dollars in damage. It hit extreme southern Florida and southeastern Louisiana and did significant damage in Miami, the Florida Keys, and New Orleans.

Storm Name or Location	Year	In Billions of 2005 Dollars (rounded up to nearest billion)
1. Miami	1926	148
2. Galveston	1900	96
3. Katrina	2005	81+
4. Galveston	1915	66
5. Andrew	1992	54
6. Long Island Express	1938	38
7. SW Florida	1944	38
8. Okeechobee	1928	33
9. Donna	1960	28
10. Camille	1969	23
11. Wilma	2005	21
12. Betsy	1965	19

*Based on work by Roger A. Pielke et al. in November 2006, updating previous studies. Values are based on an average of two methods for determining a normalized damage figure, rounded up to the nearest billion.

Seasons with the Most Storms

Year	Total Storms with Winds of at Least Tropical Storm Strength	Number of Hurricanes
1. 2005	28	15
2. 1933	21	10
3. 1887	19	11
4. 1995	19	11
5. 1969	18	12
6. 1936	16	7
7. 2003	16	7
8. 2000	15	8
9. 2001	15	9
10. 2004	15	9

Note: There is a good chance that 1933, 1887, and 1936 had more tropical cyclones of at least tropical storm strength. Without satellites to see them, short-lived storms could easily have come and gone well away from land without official notice.

Seasons with the Fewest Storms Since 1970

Year	Total Storms with Winds of at Least Tropical Storm Strength	Number of Hurricanes
1. 1983	4	3
2. 1965	6	4
3. 1977	6	5
4. 1982	6	2
5. 1985	6	4

Note: The years 1983 and 1965, while having limited tropical activity, both produced a major hurricane that did significant damage. Hurricane Alicia hit Houston in August 1983, and Hurricane Betsy hit southern Florida and southeast Louisiana in September 1965. It happened again in 1992, when Hurricane Andrew hit South Florida and Louisiana. Interestingly, all of these hurricane seasons came on the tail end of a significant El Niño.

3

Hurricane Science

Hurricane Science—The Basics

The things we call storms come, for the most part, in two flavors: *tropical cyclones* and *extratropical cyclones*. Tropical cyclones get their energy from warm ocean water. That's why they exist mostly in the tropical regions of the world, where the sea surface temperature is high. Extratropical cyclones get their energy from the interaction of warm and cold air masses. That's why they exist farther north, where cold air lives. But nothing that Mother Nature does is ever that simple. We'll deal with hybrids later.

Important Definitions

Hurricane. A tropical cyclone with sustained surface winds higher than 73 mph in the Atlantic Ocean, Caribbean Sea, Gulf of Mexico, and the North Pacific Ocean east of the International Dateline.

Sustained wind. Defined in the United States as a 1-minute average wind measured at 33 feet (10 meters) above open terrain. (See page 92.) Other averaging standards are used elsewhere in the world. Gusts are similar, but the averaging period is 3–5 seconds.

The Word *Hurricane*

Hurricane entered English via Spanish. The mythology of the Caribbean Taino Indians included a god, Juraká, who ruled over the hurricanes. Jurakán became *huracán* in Spanish. Christopher Columbus found a thriving civilization of Taino Indians when he arrived at what is now Mole St. Nicholas, Haiti, on December 6, 1492. The Taino god was apparently derived from the Mayan god, Hurakan (spelled many ways, and meaning "one-legged"), the mythological creator of the world, the animals, men, and fire. When the gods became angry with the first humans, the Mayan story goes, Hurakan unleashed the Great Flood that destroyed them. In appearance his missing leg was replaced by a serpent or a snout.

Cyclone. The generic name for any low-pressure area with a closed circulation.

Tropical storm. An organized tropical system with a defined circulation and maximum sustained winds between 39 mph and 73 mph.

Tropical depression. An organized tropical system with a defined circulation and maximum sustained winds of less than 39 mph.

Tropical wave. Also called an "easterly wave." A defined kink or bend in the trade winds blowing from east to west over the tropical Atlantic. On a map showing the direction of the wind, it appears as a hump or a wave in the flow.

Tropical disturbance. A discrete weather system of thunderstorms in the tropics not associated with a frontal system that maintains its characteristic for twenty-four hours or more. Tropical disturbances are generally the "seeds" from which tropical storms and hurricanes grow.

Units in This Book

Wind speeds. In this book I've decided to discuss wind speeds in *miles per hour* (mph) whenever possible. Unfortunately, we'll find wind

Tropical Wave or Tropical Disturbance

Very often the terms *tropical wave* and *tropical disturbance* are used interchangeably, but that is not always correct. Often a tropical wave generates a thunderstorm as it moves through the tropics. In this case, it would qualify as a tropical disturbance. Sometimes, however, a tropical wave can move through dry air and have no accompanying rain showers. In this case the wave is only identified by its effect on the trade winds and therefore is not, technically, a tropical disturbance.

reports and forecasts in various units, including knots, kilometers per hour, and meters per second. But, archaic as it is, the National Hurricane Center's Public Advisories are in "mph," and that's what most people in the United States can best relate to. So miles per hour it is. Many of the bulletins report wind speed in *knots*, however. There's much more on that below.

Pressure. Although it's mixing units systems, we'll only talk about pressure in the metric unit *millibars* (mb). Historically, a hurricane's central pressure was reported in *inches of mercury*, but we don't do that anymore. All of the discussions from the National Hurricane Center are in millibars. Occasionally you may run into *hectopascals* (hPa). But 1 hPa = 1 mb, so there's no difference.

Distance and height. Back to the imperial units. Distances will be discussed in *statute miles*. Wave heights, and so on, will be in *feet*.

Measuring the Wind

Wind is not an easy thing to measure. First of all, the moving air interacts with the buildings, the trees, the ground, and everything else, swirling and gusting and generally carrying on. It's not at all a uniform thing. It never blows evenly, smoothly, or anything like it. So, the only thing to do is take an *average reading* over time. This average is then converted, if necessary, to what the wind speed *would be* at a *standardized* location. Without standardization, there would be no way to make reasonable comparisons of the wind at different places and different times.

On land, the friction with the earth's surface disrupts and slows the wind flow. Everything else being equal, the same weather system would likely generate higher wind speeds if it were over the water. But the real world isn't full of open spaces, so in general the wind speed gets estimated or adjusted by taking obstructions, elevation, and other factors into account.

History of Measuring Wind

Mariners have been estimating how strong the wind is blowing ever since there were boats. As we have seen, estimating is not a bad idea, since measuring the wind, even with modern instruments, is not easy.

In the early nineteenth century, British Commander Francis Beaufort devised a wind scale for use in his private log.

Eventually, he settled on 13 steps for the scale, from 0 for calm winds to 12 for winds of hurricane force. The speeds would be estimated by the number and type of sails a frigate would use in each level of wind strength. For example, a wind of Beaufort 4 strength would "propel a well-conditioned man-of-war under full sail in smooth water at a speed of 5 to 6 knots." (See *knots*, page 92.) In a hurricane wind of Beaufort 12 strength, the ship would "show no canvas."

In 1838, with Beaufort now an admiral, the scale was adopted by the British Royal Navy. Through the years, because of the evolution of sailing vessels, the scale has been modified slightly so that wind-speed estimates can be made by looking at the sea state, and a system for estimating wind speeds on land was added as well. But, basically, Admiral Beaufort's scale survives.

Beaufort Wind Scale

Force	Wind (mph)	WMO Classification	Appearance of Wind Effects	
			On the Water	On Land
0	Less than 1	Calm	Sea surface smooth and mirrorlike	Calm; smoke rises vertically
1	1–3	Light Air	Scaly ripples, no foam crests	Smoke drift indicates wind direction; still wind vanes
2	4–7	Light Breeze	Small wavelets, crests glassy, no breaking	Wind felt on face, leaves rustle, vanes begin to move
3	8–12	Gentle Breeze	Large wavelets, crests begin to break, scattered whitecaps	Leaves and small twigs constantly moving, light flags extended
4	13–19	Moderate Breeze	Small waves 1–4 ft. becoming longer, numerous whitecaps	Dust, leaves, and loose paper lifted; small tree branches move
5	20–24	Fresh Breeze	Moderate waves 4–8 ft. taking longer form, many whitecaps, some spray	Small trees in leaf begin to sway

(continued)

Force	Wind (mph)	WMO Classification	Appearance of Wind Effects	
			On the Water	On Land
6	25–31	Strong Breeze	Larger waves 8–13 ft., whitecaps common, more spray	Larger tree branches moving; whistling in wires
7	32–38	Near Gale	Sea heaps up, waves 13–20 ft.; white foam streaks off breakers	Whole trees moving, resistance felt walking against wind
8	39–46	Gale or Tropical Storm	Moderately high (13–20 ft.) waves of greater length, edges of crests begin to break into spindrift, foam blown in streaks	Whole trees in motion, resistance felt walking against wind
9	47–54	Strong Gale	High waves (20 ft.), sea begins to roll; dense streaks of foam, spray may reduce visibility	Slight structural damage occurs; slate blows off roofs
10	55–63	Storm	Very high waves (20–30 ft.) with overhanging crests, sea white with densely blown foam, heavy rolling, lowered visibility	Seldom experienced on land; trees broken or uprooted; "considerable structural damage"

Force	Wind (mph)	WMO Classification	Appearance of Wind Effects	
			On the Water	On Land
11	64–73	Violent Storm	Exceptionally high (30–45 ft.) waves; foam patches cover sea; visibility more reduced	
12	74+	Hurricane	Air filled with foam; waves over 45 ft.; sea completely white with driving spray; visibility greatly reduced	

Beaufort Formula

In 1903, an official formula was devised to convert Beaufort numbers to miles per hour:

$$W \text{(mph)} = 1.87 \times \text{square root } (B^3)$$

where W is the speed of the wind and B is the Beaufort number.

Why Do Hurricanes Start at 74 MPH?

Have you ever wondered why the beginning wind speeds for tropical storms and hurricanes are not round numbers—40 mph or 75 mph, for example? Take a look at the table. It's Admiral Beaufort's fault.

Beaufort 8–11 or 39–73 mph = tropical storm force
Beaufort 12+ or 74+ mph = hurricane force

The wind-speed ranges for tropical systems break evenly on the Beaufort numbers, not a round number in *knots* or *mph*.

Knots Versus MPH

Hurricane information *should* be given in knots (kt). Airplane pilots and mariners use knots as the standard unit of speed. But in the United States people are used to miles per hour, so the National Hurricane Center converts to mph for their public advisories. Issued at the same time as the Public Advisory—which is designed to communicate the hurricane forecast in an explanatory text format—are the Forecast Advisory (a more tabular form of the forecast) and the Forecast Discussion (a somewhat technical explanation of the rationale behind the forecast). Wind speeds in the latter two bulletins are reported in knots, so wind speeds from the National Hurricane Center are always a bit confusing. To make matters worse, most countries in the world—except the United States— use kilometers per hour (kph or km/h) or meters per second (mps or m/s).

Sustained Wind Averaging Times

Most countries use a longer averaging time for "sustained" winds than the United States does. Australia, for example, uses a ten-minute average, not our one-minute average. There's a complicated conversion formula, but it works out that a 75 mph Australian storm is about 10 to 15 percent stronger than a 75 mph American storm, depending on whether it's over water or over land.

Pressure Versus Wind

Because the highest winds in a storm are so difficult to measure precisely, the lowest barometric pressure at the center of a storm is often used to gauge a hurricane's strength. It's well known, however, that two storms with the same pressure can have significantly different maximum wind speeds. Still, in general, storms with lower barometric pressure are stronger.

Two Kinds of Miles

The fact that the same wind speed in knots and mph are close to the same number is purely coincidental. Knots are *nautical miles per hour*, while mph means *statute miles per hour*.

The word *mile* comes from Roman times. If you know a Romance language—French, Spanish, Italian—you'll recognize its similarity to the word for a thousand. *Mille passum* means "1,000 paces" in Latin. A *pace* in this case means two steps. So a Roman "pace" was equivalent to about five feet, meaning a Roman mile equaled about 5,000 feet.

- A *statute mile* was defined in Britain in 1593. By order of Queen Elizabeth, a mile was increased to 5,280 feet to equal exactly eight furlongs, a commonly used measure of length from medieval times. So the exact length of a mile was decreed by the queen. Thus, a *statute mile* was set by statute.
- A *nautical mile* is defined as one minute (1/60th) of one degree of a great circle of the earth that passes through both the North and South poles.

    One minute is 1/60th of a degee.

    One degree is 1/360th of a circle.

    So one nautical mile is $1/60 \times 1/360 = 1/21,600$th of the distance around the earth on a great circle through the poles. (The earth is slightly squashed—so the distance around the equator is a bit greater.)

    The legal definition is 6076.115 feet, or 1.151 times a statue mile.

Miles in This Book
In this book, *miles* means *statute miles*. *Nautical miles* will be written out explicitly.

Knots from the Sea

For hundreds of years, ships measured their speed by tying a rope to a wooden weight and tossing it off the back into the water. Knots were tied in the rope every 50 feet or so. Using a 30-second hourglass, sailors would time how many knots were pulled off the ship in that time. The more knots, the faster the ship was going.

The distance between knots was calibrated for easy math. At 50 feet every 30 seconds, that's 100 feet every minute, or 6,000 feet every hour; 6,000 feet is about a nautical mile. So one knot pulled off the stern every 30 seconds is one nautical mile per hour.

Today, that's the official definition:

1 knot = 1 nautical mile per hour = 1.151 statute miles per hour

Knots are only used over the water and in the air. That's why ships, buoys, radars, satellites, and Hurricane Hunters all report the winds they measure in knots. The speeds are converted to mph for the public advisories, but left in knots for the other NHC bulletins. (See Understanding Hurricane Advisories," page 154.)

Converting 115 Knots to MPH

There's a wrinkle in the conversion of knots to mph in NHC Public Advisories. Normally, the rule is: Multiply knots times 1.151 and round the result to the nearest 5 mph. For example:

90 knots × 1.151 = 103.59

Therefore, the public advisory would say 105 mph. But:

115 knots × 1.151 = 132.365

That would normally be rounded down to 130 mph. But because the Saffir-Simpson scale defines a 115-knot hurricane as category 3, the wind speed is rounded *up* to 135 mph, so that it falls into the category 3 range in mph as well. (See "Saffir-Simpson Hurricane Scale," page 148.)

Current and forecast wind speeds are only reported in 5-knot or 5-mph increments in recognition of the uncertainty intrinsic in the wind-speed values.

Milestones in Hurricane Science

Imagine a time before satellites. You know a hurricane is moving toward the coast, but you can't see it. If you are much over forty, you lived in the era when hurricanes were "invisible" until they were just offshore. If you are over seventy, you lived in an era when the "steering currents" in the upper atmosphere were mostly unknown.

Imagine living in a time when you had no concept of what a storm, any storm, was. Is a storm an identifiable entity? Or does weather just happen? Without long-distance communications there was no way to correlate weather observations and visualize a weather pattern, or even imagine there was such a thing as a weather pattern.

Many people contributed, of course, to our understanding of weather, and specifically of hurricanes. As with so many things, it just took wide eyes and an open mind, and a willingness to look and to think, to make a profound discovery. In the eighteenth century one of the keenest observers in the world was living in Philadelphia.

October 22, 1743: Thick clouds that evening keep Benjamin Franklin from seeing a lunar eclipse that he'd planned to observe. The night was rainy with a strong wind blowing from the northeast. At the time, the assumption was that bad weather comes from the same direction as the wind. On this night, Franklin was sure that his brother in Boston had experienced the bad weather before him because Boston is northeast of Philadelphia. But, newspaper reports arriving in Philadelphia in the subsequent weeks convinced Franklin that he was wrong. The eclipse *was visible* in Boston. The storm (now known to be a hurricane) arrived there later that night. Inquiries of travelers confirmed that the bad weather occurred first in South Carolina and *moved* up the coast. Franklin was the first person to deduce that storms were independent entities.

1747: Lewis Evans publishes a map containing "Franklin's Rule" that "all great storms begin to the leeward." *Leeward* in this instance means in the direction the wind is blowing (as opposed to the direction the wind is coming from).

1781: A German meteorological society accumulates all available records from U.S. observers with the idea of finding trends that might lead to long-range forecasts.

1801: James Capper, an amateur meteorologist from Wales working in India, proposes that storms in the Indian Ocean were "whirlwinds whose diameter cannot be more than 120 miles." It's not clear that this idea was widely distributed, however.

1819: During the period from 1807 to 1817, Professor John Farrar of Harvard University maintains a full weather record at Cambridge, Massachusetts. After the Great September Gale of 1815 (see page 36), he theorizes that the storm "appears to have been a moving vortex and not the rushing forward of a great body of the atmosphere." He expounds on this idea that storms have rotation in a paper in 1819.

July 1831: William C. Redfield publishes a groundbreaking article in the *American Journal of Science and Arts* detailing his theory that the New England hurricane of 1821 (see page 37) was a "great whirlwind" that had originated in the West Indies (the Caribbean Islands). Traveling around Connecticut, he documented that the trees fell with different orientations during the storm. The only explanation was that the winds rotated around a center. The article also proposed the idea of steering currents, speculating that storms move "as a component portion of the general mass of atmosphere which has previously been tending in that direction." And he was the first to *track* a hurricane when he chronicled the movement of the August 1830 hurricane from the Virgin Islands along the east coast to New York. Redfield's article was the first reasonably accurate view of the surface structure and the movement of hurricanes.

1838: Lieutenant Colonel William Reid is sent by the British government to Barbados after a major hurricane there in 1831 to help with reconstruction. He finds that Redfield's theories fit the evidence he finds in the ships' logs from the Great Hurricane of 1780 (see page 36). Reid presents a paper in 1838 expounding on Redfield's theories and setting out his own "Law of Storms." He publishes a book three years later of that name, which becomes a standard set of rules for mariners to avoid storms.

1847: The same William Reid, now the British governor of Barbados, establishes the first system for displaying warning signals when the approach of a storm is indicated by the barometer.

1866: A Jesuit priest, Father Federico Faura, joins the Manila (Philippines) Observatory and begins a study of meteorology and astronomy.

1870: Father Benito Viñes, also of the Jesuit order, becomes director of Belen College in Havana and begins studying hurricanes with the goal of establishing a hurricane warning system. The Jesuits had a long tradition of scientific research, but he might have been motivated by the two hurricanes that hit the Havana area that year. The two Jesuit priests on opposite sides of the globe would become the world's experts on hurricanes.

1870: Congress gives the Army Signal Corps responsibility for organizing a national meteorological service in the United States.

1871: Cleveland Abbe joins the Corps as its first chief meteorologist. He brings scientific reasoning to forecasts, where previously they were just "generalizations based upon observations." Forecasters are called indications officers.

August 6, 1873: The Signal Corps gets its first weather reports from Cuba. Reports from other Caribbean islands soon follow. Drawing weather maps of the tropical region is now possible.

September 28, 1874: The first printed weather map generated by the Signal Corps that shows a hurricane. The storm is offshore between Jacksonville and Savannah. (See following page.)

September 11, 1875: Father Viñes issues the first known official hurricane warning two days before a hurricane devastates the southern Cuban coast.

First plotted hurricane (indicated as a LOW) on a U.S. weather map depicting the weather pattern at 7:35 A.M., September 28, 1874. Courtesy NOAA Central Library Data Imaging Project.

It is possible he did it before that date as well. Hurricanes hit Cuba in 1873 and 1874, but there is no record of any advisories. Viñes makes the predictions with advanced, for the time, scientific analysis. He reviews weather data from a network he has set up across the island. In addition, cloud movements at various levels in the atmosphere are carefully examined, and changes in the sea swells are noted.

1876–1881: Caribbean reports to the United States are suspended, reinstated, and then suspended again. U.S. government funding is cut off due to a complicated fight in Congress over whether it is legal for the federal government to spend money outside the country, as well as other issues in the military budget. The amount they were fighting over was $3,000.

1876: Father Viñes issues another "successful" hurricane warning. He is praised throughout Cuba and his reputation grows internationally.

1889: The Cuban Meteorological Service is officially organized and sends reports from throughout the Caribbean to the United States.

July 1, 1891: United States Weather Bureau begins operations under the Agriculture Department.

1898: With the outbreak of the Spanish-American War, the U.S. government funds the deployment of American weather observers in the Caribbean, replacing the Cuban network. President McKinley says he fears a hurricane more than the Spanish navy. Kingston, Jamaica, is designated the head office of the new Caribbean hurricane-warning service.

1899: With the end of the war and Americans in control in Cuba, the headquarters of the Caribbean-area hurricane-warning service is transferred to Havana. Responsibility for hurricanes near the United States remains in Washington.

Around 1900: The Weather Bureau experiments with using kites to get upper-air data.

September 8, 1900: The deadliest hurricane in American history hits Galveston, Texas. No hurricane warnings are issued in spite of regular communiqués sent to Washington by the head of the Weather Bureau in Galveston, Isaac Cline. See page 42.

1902: Responsibility for Caribbean hurricane warnings is transferred to Washington when the American troops leave Cuba. Hurricane forecasts only extend twelve to twenty-four hours and are mostly based on the current path of the storm.

1909: Weather balloons are used to get some upper-air data.

August 26, 1909: The first-ever ship report of a hurricane reaches the Weather Bureau.

June 1, 1919: A hurricane forecast center opens in San Juan with responsibility for the eastern Caribbean and the tropical-Atlantic region.

1931: The kite observation system is discontinued because the kites are hazards to aviation.

1934: A system for soliciting information from ships is initiated.

July 1, 1935: The hurricane-warning system is dramatically upgraded. Responsibility for hurricane forecasting is divided among forecast centers in Washington, San Juan, Jacksonville, Florida, and New Orleans. For the first time Teletypes connect these locations with other local Weather Bureau offices.

1937: A sophisticated system of gathering data in the upper atmosphere is initiated by sending instrument packages—called radiosondes—aloft in balloons. Pan American World Airways, the U.S. flag carrier, gathers weather information at its flight terminals and forwards the data to the Weather Bureau. For the first time forecasters are able to analyze steering currents.

1940: President Roosevelt transfers the Weather Bureau to the Department of Commerce.

1943: A primary hurricane forecast office is established in Miami, replacing the Jacksonville office. Grady Norton is named the head of the new office. He is considered to have been the first director of the National Hurricane Center, although the position wasn't called that at the time. The office is co-located with the Miami Weather Bureau.

July 27, 1943: Colonel Joseph Duckworth becomes the first (confirmed) person to intentionally fly a plane into a hurricane's eye. An accomplished pilot who taught classes on how to fly into bad weather in the Army Air Force, Duckworth twice flew into a hurricane near Houston in a single-engine AT-6 Texan training plane. Legend says it was done on a dare. See page 53.

1944: Regular flights into storms begin because of the losses being taken by the American fleet from bad weather. The first Weather Reconnaissance Squadron is based at Gander, Newfoundland, to support troop and supply movements to Europe.

Mid-1940s: Radar is used to track weather systems for the first time. Patented in 1935 and deployed as a wartime technology to locate enemy aircraft, radar was found to have an application as a weather-tracking tool as well.

October 9, 1954: Chief forecaster Grady Norton dies during Hurricane Hazel (see page 61). Gordon Dunn from Chicago replaces him in 1955.

1955: A new radar at Cape Hatteras "sees" all three east coast hurricanes that year. Because of the six-storm hurricane barrage of 1954 and 1955 (see page 59), Congress appropriates money for a national hurricane research project and the deployment of a coastal radar system. By 1960, the radar "fence" along the coast is up and running.

September 26, 1955: A navy reconnaissance plane goes down in Hurricane Janet in the Caribbean, killing nine navy men and two Canadian journalists.

1956: The National Hurricane Research Project begins. Based in West Palm Beach with three aircraft, researchers fly into every available storm to gather data.

Late 1950s: Early computer-forecast models are run.

April 1, 1960: The first American weather satellite, Tiros I, snaps its first picture. The satellite only stayed in orbit seventy-eight days, but it proved the value of satellite coverage of the weather.

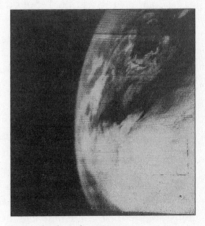

August 23, 1963: The Project Storm Fury experiment to try to weaken hurricanes by "seeding" them with silver iodide officially begins. Ten aircraft are sent into Hurricane Beulah north of Puerto Rico. The complicated mission was an operational success, but after seeding a handful of storms,

Tiros I had a television camera on board. This was man's first live view of the earth from space.

scientists could never determine if they had any impact on the strength of the storm, or if the fluctuations in intensity were natural. The project is finally closed down in 1983.

1965: Five computer models are in regular use for hurricane forecasting, all based on statistics. Models are developed in conjunction with the National Hurricane Research Project.

February 1966: NASA launches the ESSA-1 and ESSA-2 weather satellites. Forecasters for the first time have regular access to the view from above. Continuous images were not available, but every spot on the earth was photographed each day, so no longer could a storm surprise forecasters and show up at the coast unannounced.

1970: The Weather Bureau becomes the National Weather Service, a part of the new agency called NOAA, the National Oceanic and Atmospheric Administration.

1970: A computer-forecast model called SANBAR, which simulated a simple version of the atmosphere, is used with some success.

October 16, 1975: NASA launches GOES-1, the first weather satellite to be "parked" over a fixed spot on the earth. Forecasters have continuous satellite coverage for the first time. GOES stands for Geostationary Operational Environmental Satellite.

1976: NOAA acquires its first P-3 aircraft for research and reconnaissance. Another one comes in 1977.

1977: A direct satellite link between reconnaissance aircraft and the National Hurricane Center is operational. Now forecasters can know what is going on inside the storm in real time.

Early 1990s: The new generation of radars is deployed by the National Weather Service. Dubbed NEXRAD, these Doppler radars have longer range, higher resolution, and can measure wind speeds when storms are within about seventy-five miles of the coast.

1992: An advanced hurricane computer model from the Geophysical Fluid Dynamics Laboratory (called the GFDL) at Princeton is put into operation. Computer-generated forecast tracks improve.

April 13, 1994: The next generation of GOES satellites is operational, giving forecasters higher resolution and more frequent images.

1997 hurricane season: "GPS dropwindsondes" are used for the first time. The high-tech instrument packages are dropped from the new high-flying NOAA jet to make precise readings from flight level down to the surface. The new data dramatically improves hurricane forecasts.

During Hurricane Season

Getting a Tropical Cyclone Started

Hurricanes and tropical storms need a "seed" to get started. Generally, there are two types of systems that can serve as the seed:

- Tropical waves, which move west from Africa
- Extratropical (nontropical) systems, which evolve over time to take on tropical characteristics

Tropical waves are the seeds for about 60 percent of the tropical storms and hurricanes that develop, but they account for about 85 percent of the major, most destructive hurricanes.

Tropical waves. Large thunderstorm complexes develop over the African land mass and travel westward in the trade-wind belt across the eastern Atlantic. Sixty to a hundred of these systems move through the tropics every year, but most have unremarkable lives.

The systems are called "waves" because they travel through the wind flow much as a wave travels through water. The air bunches up, and that forces some moisture to rise in the atmosphere, creating clouds and sometimes rain. Very occasionally, however, a favorable weather pattern meets up with one of these systems and enhances the lifting of the moist air.

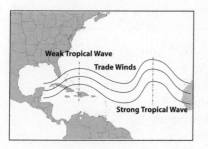

As the lifting is enhanced, more air gets drawn into the system at the surface. Air converging on the wave begins to turn, due to the spinning of the earth (the Coriolis force), and the system begins to rotate. If all of the necessary components are in place, a tropical cyclone develops.

It doesn't happen very often because all of the ingredients in the wave, the atmosphere, and the ocean aren't usually present at the same time. The wave has to be in the right place, at the right time, with the right attributes. Here are the key issues:

1. The atmosphere needs to have enough cool air above the surface to provide a contrast with the ocean-warmed air. Warm air rises faster with cooler air above it.
2. The water temperature should be at least 80°F.
3. The air above the surface needs to be moist. On occasion, dry air from Africa moves out over the Atlantic, inhibiting tropical development.
4. The atmospheric pattern must be favorable. Winds must be blowing about the same speed and direction at all levels of the atmosphere; that is, there has to be low wind shear. If there are strong winds aloft, the tops of developing cells get blown off, inhibiting development.
5. The system needs to have maintained some of the internal rotation it had over Africa.
6. The wave needs to be far enough north of the equator for the Coriolis effect to turn the air that is being pulled in. There is no Coriolis effect at the equator.

East Atlantic tropical waves are responsible for most of the major hurricanes—although the 2005 season was an exception—because storms that form well away from land have time to organize, grow, and strengthen before reaching land.

In this discussion we take the word "atmosphere" to mean the part of the atmosphere closest to the earth, which is where the storms are. Technically, the atmosphere extends much higher, but those upper levels are not important to us. The lower part of the atmosphere is properly called the troposphere.

Other Tropical Triggers

Just as the African thunderstorm complexes are the initiating entities for tropical waves, fronts and other extratropical systems can initiate tropical disturbances as well. If one of these systems is going to trigger tropical development, however, it usually happens early or late in the hurricane season, when "northern" weather systems can move closer to the tropical latitudes.

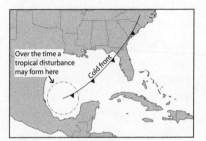

A common weather pattern in the fall. Fronts that stall over warm water are sometimes the trigger for tropical cyclones after the weather pattern aloft changes to one that is favorable for tropical development.

Fronts as triggers. For example, if a front moves south and stalls in the Gulf of Mexico, it normally brings along cool, dry air that initially inhibits any tropical development. Over time, however, the waters of the Gulf warm and moisten the air. The front can then become the triggering disturbance if the weather pattern in the overall atmosphere is conducive to it. An extratropical system normally has to "sit" for a while so that the air in the system can warm and become more humid. As a result, the Tropical Weather Outlook from the National Hurricane Center often says something like "Any development will be slow to occur."

Upper-level low as a trigger. On occasion, an upper-level extratropical low pressure system that moves out over the warm ocean can be the trigger. The low can generate thunderstorms. The air that rises in those thunderstorm cells is warm and moist, which in turn warms the surrounding atmosphere. Sometimes, over time, a tropical cyclone forms.

Tropical cyclones sometimes form from these extratropical systems over water that is slightly cooler than 80°F. This is mostly true when the trigger is a low-pressure system with a healthy circulation already.

2005 Storm Development

The mega storms of the 2005 hurricane season—Katrina, Rita, and Wilma—all formed from tropical waves, as you might expect. But none of them formed in the Atlantic east of the Caribbean, where most major hurricanes are born.

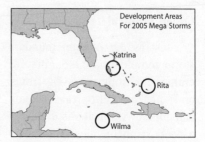

The atmospheric environment over the Gulf of Mexico, south-western Atlantic, and the Caribbean was exceptionally conducive for tropical development in 2005. In addition, the water temperatures ranged from slightly above normal in the Gulf to more than 2°F above normal in parts of the Caribbean. Still, it's not clear what was "wrong" with the normal breeding grounds in the tropical Atlantic.

The record-breaking storms of 2005 all formed close to the U.S. coastline. Historically, most large and powerful hurricanes have developed from tropical waves far to the east, where they have time to grow and strengthen.

What Makes Hurricane Season Special?

During the winter months, the overall weather pattern around the earth shifts south as the southern hemisphere warms and the northern hemisphere cools. The polar jet stream dips across North America and the Atlantic Ocean. This fast-moving high-altitude stream is the southern boundary of cold, northern air.

Fast-moving air in the upper atmosphere will not allow tropical development. In addition, the ocean temperature cools too much over most of the areas that are favorable for development in the summer.

The jet steam is always moving, oscillating north and south. That's why it's not cold every day in the winter, even in the northern states. On a rare occasion in the winter months, when the jet steam temporarily retreats to the north, a disturbance can come along over one of the areas where the water is still reasonably warm and tropical development can take place. Most often the triggering disturbance starts out as an extratropical storm. The system sits

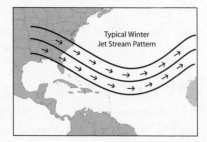

over an area of warm water and transforms. These systems usually don't last long because they soon move over too-cool water or the weather pattern changes.

Subtropical Depressions and Storms

Subtropical Storm Ana on April 20, 2003, shortly before it changed to tropical storm status. The first tropical storm on record to form in April. Courtesy NASA.

Hurricanes and tropical storms are examples of *tropical cyclones*; a nor'easter is an example of an *extratropical cyclone*. And, there's a gray area between the two. A storm system that has some attributes of each classification is called a *subtropical cyclone*.

Subtropical cyclones are divided as you might expect:

* Subtropical depression—Maximum winds less than 39 mph
* Subtropical storm—Maximum winds over 39 mph

There is no *subtropical* category comparable to a hurricane. Only one subtropical storm in the record book, the last storm of the year 1979, briefly reached hurricane strength. It was quickly absorbed by a large extratropical low over northeastern Canada.

The National Hurricane Center bases its classification of a storm on several factors. Often it has to do with the history of the system. Ana of 2003, for example, was initially an extratropical low that broke away from a front between Bermuda and the United States. When a front is involved in the "birth" of the system, the assumption is, at least initially, that it is probably subtropical.

The distinguishing features of a subtropical cyclone are:

* Often the system was previously part of a frontal system or an extratropical storm.
* Strong winds occur in a wide zone up to 120 miles from the center.
* The temperature of the air aloft in the core of the storm is cool.

The initial advisory indicated that the strongest winds in Ana extended as far out as 85 miles from the center. Based on this, Ana was designated a subtropical storm. In a later analysis, satellite temperature measurements indicated that the upper-level core had warmed by the evening of April 20 sufficiently to designate Ana a tropical storm. While the storm was active, that change was not made until the next day. Top winds from late on April 20 to April 21 were determined to have been 60 mph. Ana once again became an extratropical storm as it merged with a cold front three days later.

El Niño and La Niña

Peruvian Fisherman and El Niño

Periodically, fishermen on Peru's west coast notice a significant warming of the temperature of the Pacific Ocean, which diminishes the fish available to catch. Most years a cold current runs north along the South American coast (called the Humboldt Current), much as the California Current brings cold water from the north along the west coast of North America.

The Humboldt Current turns left (away from Peru and Ecuador) as it nears the equatorial region, partially due to the spinning of the earth. This flow away from land allows nutrient-rich water from deep in the ocean to rise to the surface. The phenomenon is called upwelling. Under this oceanic flow regime, the weather in coastal Peru is normally quite dry and the fish are plentiful.

On occasion, however, everything reverses. A warm current comes from the north, the weather is more humid, flooding rains arrive, and the fish stay away. Some times these events last only a month or two, other times they last for many months or a year. Long episodes are extremely disruptive to the livelihoods of the fishermen and the economy of the area.

Because the warm-water events often occurred around Christmas, they came to be called El Niño (Spanish for "the Christ child").

1982–83 El Niño Gets People Interested

The El Niño phenomenon was identified and quasi-officially named in the late nineteenth century. But it was the devastating effects of the extremely strong El Niño in 1982 and 1983 that got scientists interested in trying to figure out what was going on.

Worldwide weather patterns were dramatically changed by the exceptionally warm pool of Pacific water west of South America near the equator. From torrential rain and landslides in California to deadly droughts in Australia and Africa, every continent was affected.

In the early twentieth century, a British scientist named Sir Gilbert Walker discovered that El Niño seemed to be related to the atmospheric pressures across the Pacific. This was a breakthrough linking the atmosphere and the ocean, and it implied that changing wind patterns were somehow related to El Niño.

Air moves (wind blows) from high pressure to low pressure. *The bigger the pressure difference, the stronger the wind.* So, if the high pressure becomes weaker, the pressure difference is less and the wind relaxes. Was Niño affecting the pressure patterns or vice versa? In either case, that early observation was important to our understanding of the combined ocean and atmosphere system.

Modern research gives us a reasonable picture of how El Niño affects the weather, but how these events get started in the first place is still not clear.

Understanding El Niño

We now know that the atmosphere and the oceans are really one system. Weather patterns affect the oceans, and changes in ocean currents or sea-surface temperatures affect the weather. This makes determining cause and effect difficult, if not impossible.

Since the system is something of a circle, we have to pick a place to start. So let's choose the wind. Normally trade winds are relatively strong and enhance the effect described above. The wind blowing toward the west across the tropical Pacific pushes surface water to the west away from the South American coast, allowing upwelling, so the water close to land is cool.

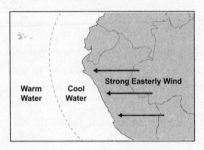

When the trade winds from the east are weak, however, the surface water is *not* pushed off-shore, the cold water stays below the surface, a weak but warm current flows from the north, and coastal waters heat up. This is an El Niño condition.

The changing currents drive the fish away and have a dramatic affect on the weather. Here's the rest of the circle. During an El Niño, the warm water near the coast heats the air directly above it, and, of course, warm air rises. Rising air lowers the atmospheric pressure (actually it weakens the high pressure). As noted above, lower pressure means a weaker wind. Less wind means more warm water, which means lower pressure yet. So once things get started, they can be self-reinforcing.

The "starting" mechanism is elusive, however. El Niño returns every two to ten years, according to the historical record. Apparently, many factors are involved, including various waves of energy that are known to propagate around the earth that make the timing and the intensity quite variable from one El Niño to the next.

ENSO: El Niño and the Southern Oscillation

Around the turn of the twentieth century, Sir Gilbert Walker closely studied the weather data from the Pacific and Indian oceans. He found that periodically the differences in atmospheric pressure between locations in the east and the west "oscillated." That is, sometimes there was a lot of difference, other times not. Sometimes the east was higher, sometimes the west. He called this phenomenon the Southern Oscillation.

Now we know that pressure changes, wind flow, and water temperatures are all interrelated. It's just not clear which of the players, if any, is leading the dance.

ENSO, for El Niño–Southern Oscillation, is the term for the whole set of interrelated phenomena. El Niño and ENSO are often used interchangeably, which is incorrect. El Niño is only the warm-water phase of ENSO. The cold-water phase is called La Niña.

El Niño and Hurricanes

When the water is unusually warm in the eastern tropical Pacific (the El Niño effect), hurricane activity in the Atlantic is diminished. Major hurricanes (categories 3–5) are especially affected. Research has shown that there are about one-third as many landfalling major hurricanes on the U.S. coast during El Niño years as when the water is cool. In addition, two landfalling hurricanes are about half as likely to occur during an El Niño as when the water temperature off South America is normal.

The reason this happens is fairly straightforward. As noted above, during El Niño events more air than normal is rising in the atmosphere over the eastern tropical Pacific. That extra air has to go somewhere, and some of it spreads out through the upper levels of the atmosphere, affecting the winds that blow across the tropical Atlantic. This air pushing from the west (from the Pacific) blows the tops off tropical systems as they try to form, thus limiting both the number of hurricanes that develop and diminishing their intensity.

La Niña and Hurricanes

On occasion, strong easterly winds push the warm water far away from the South American coast near the equator. Exceptionally cold water then rises from the lower depths of the Pacific. When this cold-water condition persists for six months, a La Niña is said to be under way.

The Atlantic weather pattern is affected in the opposite way from an El Niño. Air above the tropical Pacific is cooled. It is therefore heavier, and it sinks. This downward "suction" pulls air from the upper atmosphere, which enhances the wind from the east above the tropical Atlantic, a condition favorable for hurricane development.

A Few Technical Details

The explanation above is a simplified version of ENSO. Notice, I only briefly mentioned the spinning of the earth, which also has a significant effect on how the warm and cold water move. And, from the paragraphs "El Niño and Hurricanes" you might have the impression that El Niños create a wind blowing from west to east across the Atlantic. Normally, that doesn't happen. The spinning of the earth, which creates the easterly winds over the Atlantic, is the stronger force. But the El Niño effect reduces that easterly flow so the air in the upper atmosphere is moving relatively slowly. This creates a *nonuniform* flow vertically in the atmosphere that is not conducive to tropical development.

As you can see, the ocean/atmosphere system is extremely complex, and not completely understood. During an El Niño both the winds and the ocean currents are affected. Each change in the atmosphere is linked to a corresponding change in the currents and vice versa. Our lack of understanding of the triggering mechanism contributed to forecasters' missing the development of a storm-limiting El Niño during the 2006 hurricane season.

Global Warming and Hurricanes

The Earth Is Warming

The earth is warming, and the rate of warming is increasing. If the trend continues, this will be a different planet in the lifetimes of our grandchildren. Here are just some of the changes that may occur:

- Animal species will disappear.
- Sea level will rise because of glaciers melting and warm-water expansion.
- Weather patterns will change (some areas being warmer and some colder).
- Diseases may increase and spread.
- Farming will likely be possible farther north in parts of the world and for more of the year, which may increase food production.

That the earth is warming is undeniable. However, the causes and effects of the warming are still being hotly debated.

- Will the warming continue at the current accelerated pace? Or, will the rate of warming slow, with perhaps some periods of cooling thrown in delaying the worst effects?
- To what degree is man causing the warming?

These are the big questions with answers that the scientific community has yet to settle, so let's not deal with them here. (After all, this is a book about hurricanes.) Some factors are certain, however, and we need to understand them:

- The amount of carbon dioxide in the atmosphere is increasing. The science that says that increased CO_2 will warm the planet is valid.
- The earth, overall, has warmed about 1°F since 1970. "Not a big deal," you say? Potentially it is. Read on.
- An annoying property of ice is that it has a fixed melting point. At one magic temperature the ice disappears. In the case of arctic ice, the melting point is about 29°F. That means if the temperature of the ice goes from 28.5°F to 29.5°F, it turns to water. That may seem obvious, but the important point is that dramatic changes can happen with what, at first glance, seems like a small temperature change.
- When land or water is covered by snow and ice, most of the heat from the sun is reflected back into space. When the snow and ice melt, the land and water absorb heat, which can cause more melting.

There are other significant factors in the global-warming discussion, but the important message out of just these four is that there may be a *tipping point*, a point of no return. So, while human-induced warming might not do much in and of itself, it could start processes that feed back on themselves that would change the climate dramatically. In my opinion, this is the real concern.

Hurricanes in a Warmer World

Some recent high-profile research suggests that the outbreak of category 5 hurricanes in the last few years is related to global warming, and, even more interesting, to human factors. I don't buy it. I'm not say-ing that global warming might not have an effect on hurricanes, perhaps even make them a bit stronger, but I'm not persuaded that there is evi-dence that the monstrous destructive storms we've seen lately are in any way connected to global temperature rise.

Here are some of the relevant facts as I see them:

- In 2005, the tropical Atlantic was significantly warmer than nor-mal, perhaps about 2°F. Some places less, some places more. The water cooled significantly in 2006, but was still well above the long-term average. These short-term temperature cycles embed-ded within other multidecadal cycles have been going on forever.
- About 1°F of the increase in temperature appears to be caused by a cycle of warming and cooling that naturally occurs. Indeed, more hurricanes are occurring now during the current warm phase. But the historical record shows that a correlation between the water temperature and hurricane frequency has always been there.
- The historical record on hurricane strength, frequency, and track has to be viewed with some skepticism. Continuous satellite coverage of the tropics only began in 1975. By that time, the downturn in hurricane activity (the Atlantic was relatively cool) had already begun. So we only have most of one downturn and part of one upturn (which we're experiencing now) with a rea-sonably complete set of satellite data.
- It's only been since 1997 that we've had a good enough under-standing of the variability in the structure of hurricanes to make informed estimates of what the winds are on the ground based on what the Hurricane Hunters are measuring at 10,000 feet, or even at 1,500 feet. Therefore, maximum wind estimates for historical hurricanes are suspect.
- Looking at the world as a whole, neither total storm activity nor intense storm activity has increased as you might expect if a warmer ocean were the only driver of hurricane strength. It is certainly possible that global warming is making a slight contri-bution to the strength of hurricanes. But, it doesn't seem to be measurable, and it appears more likely to me that other factors

(mostly cyclical atmospheric patterns) have the greatest influence over the strength and frequency of hurricanes.

- The worldwide hurricane/typhoon/cyclone record that is often cited in the studies that claim to draw a correlation between global warming and major-hurricane frequency has some glaring errors. For example, the deadliest tropical cyclone in history—hundreds of thousands of people died in Bangladesh in 1970—is not even counted. No official wind measurements were made, even though it is thought to have been an extremely strong storm.

The Thermohaline Circulation

A potentially important issue that is often discussed in the global-warming hurricane debate is the thermohaline circulation. (*Thermo* for "temperature" and *haline* for "ocean saltiness.") This worldwide conveyor belt of water, flowing at times on the surface and at times deep underwater, connects all of the oceans around the globe. The most important part of the circulation for our discussion is the Gulf Stream, which runs from Florida to the North Atlantic.

The ocean water flows north, driven mostly by the south-to-north predominant wind flow over the western Atlantic. The wind passing over the water causes evaporation, which has two important simultaneous effects: The water becomes saltier as the freshwater evaporates, and the water is cooled by the evaporative effect of the wind.

The circulation speed goes through cycles lasting twenty to fifty years. These cycles are related to natural cycles in the saltiness of the ocean. Salty water is heavier, and the saltier it is, the faster it sinks in the cold North Atlantic, making the circulation go faster.

When the system is moving fast, more warm water is transported north, and overall, the Atlantic warms up. A slow cycle yields a cooler ocean. It is thought that these changes may be part of the reason that ocean temperatures go through cycles, which may, indeed, be one of the big reasons that hurricanes go through cycles as well.

In a warmer world, with melting glaciers, the far North Atlantic would theoretically be less salty because of the additional freshwater. (Remember, the saltier the water, the faster the belt moves.) This would imply a slower circulation, and that would cool the ocean because less heat would be transported north. A cooler ocean normally means fewer hurricanes.

How would this cooling effect counteract the warming effects? Does it hold off the tipping point? In what unanticipated ways might the climate be disrupted? How do changes in the thermohaline circulation affect the frequency of El Niños and La Niños? These are all good questions having no good answers.

The Butterfly and the Bottom Line

The discussion above is just the tip of the iceberg—albeit a melting iceberg. The atmospheric/oceanographic system that is the earth's climate is so insanely complex that we may never understand all of the interactions.

The idea that a butterfly flapping its wings could trigger a violent hurricane is valid. It's just like dominoes. Imagine a room full of them. If the first one doesn't fall, they could all stand there until the cows come home. But, one little flip of a finger and the dominoes' world is changed forever. In a complex system, a small effect has the potential to be magnified. And, since small effects may not be detectable, significant events may never be forecast.

The bottom line is, nobody can say for sure exactly what the effects of global warming will be on our grandchildren's or their children's lives. The consensus, however, is that it will be somewhere between mostly manageable and very bad. Not good odds by any reasonable measure.

Given the risk, an endeavor on the scale of the Manhattan Project should be under way to deal with the threat, even if we don't understand every nuance of the climate system. The people in charge in Washington at the moment stood by and watched as a well-predicted catastrophe happened right before their eyes in Louisiana. Frighteningly, that same attitude seems to be driving the U.S. reaction to potential life-altering climate change.

Government's first responsibility is to protect its citizens. With that mission in mind, in my opinion, the United States should be leading the world in developing a technological response to the global-warming threat. That no leadership in that direction is forthcoming from Washington is unforgivable.

Storm Surge

If you live near the water, you need to understand storm surge. It is the most dangerous part of a hurricane.

- It was Katrina's storm surge that breached the levees in New Orleans.
- It was Katrina's storm surge, not the wind, that did the tremendous damage on the Mississippi coast.
- It was the storm surge, not the wind, that killed thousands of people in the great historic hurricanes.

How High Will It Get?

As a hurricane approaches land, ocean water is pushed ashore with a tremendous amount of energy, more energy than you can imagine. A one-foot-deep storm surge will sweep your car off the road. It's likely you could not stand up in a six-inch surge.

The depth of the water varies according to a variety of factors, including:

- The strength of the storm
- The angle of approach of the storm
- The shape of the coastline
- The slope and depth of the offshore seafloor

A storm surge approaching 30 feet hit the Mississippi coast during Katrina. When you consider that every cubic yard of saltwater weighs 1,728 pounds—almost a ton—you can imagine the force that is applied to structures along the coastline when that water gets moving. The surge pushed six miles inland, and double that up waterways opening to the Gulf.

A computer program run at the National Hurricane Center called the SLOSH model is reasonably good at estimating how high a storm surge is going to be as a storm approaches the coast. Based on these projections, evacuations are ordered to get people away from the rushing water. See "Evacuation Decision Making," page 207.

Water-Level Factors

What people focus on, of course, is how high the water rises. Very often that total water height is called the storm surge, but that's not technically correct. The water level is determined by four factors:

1. **Moving water.** A mound of water is pushed along with the storm a hurricane's strong winds. If you blow on a full cup of coffee, you will see that the liquid is pushed higher on the downwind side of the cup. A hurricane does the same thing. It pushes the water along as it moves forward. As the seafloor rises near the coast, this water is pushed up and over the shoreline.

2. **Low-pressure suction.** Because the atmospheric pressure in a hurricane is significantly lower than the pressure outside the storm, the water level rises near the storm's center. This is the same effect as sucking on a straw to raise water out of a glass. You are creating low pressure in the straw that causes the liquid to rise. This effect is normally small compared to the mound of water the wind pushes ashore, but in a strong hurricane it can be significant. The low-pressure effect is about one foot for each 34-millibar drop in pressure.

3. **Wave action.** A hurricane's wind puts energy into the water that generates large waves near the coast. There are two parts to this process. As the waves break, water is left on the shore because the water can't run back out; too much wind and water is pushing in. This leftover water is called "wave setup." It adds to the increased water level caused by the storm surge. In addition, the waves themselves add additional height to the water. They break on top of the water level produced by the other factors.

4. **Astronomical tides.** The normal cycles of high and low tides continue, unaffected by a hurricane. High tides add to the total

water level, and low tides, of course, subtract from it. The swing between high and low tides varies tremendously from place to place on the coastline. In some places the tides won't make much difference, and in other places they can cause a devastating flood.

Storm Surge Versus Storm Tide

- *Storm surge* is defined as the total effects from numbers 1 and 2 only. The *additional* rise in the water is due to the storm's wind and low pressure.
- *Storm tide* is defined as the total effects from numbers 1, 2, and 4. It is the storm surge and the contribution from the normal tide.

Beware of Storm Surge Numbers in Advisories

National Hurricane Center advisories give an expected storm-surge range, perhaps 8 to 12 feet. On a complex coastline, however, the range might be quite large. In Wilma the forecast was for 9 to 17 feet on Florida's west coast. Remember, however, that the water will go higher than that because of wave action and, possibly, astronomical tides. In places where the tides levels vary significantly between high and low, the time of a storm's arrival is critical to how high the water rises.

Rapid Intensification

On occasion, for reasons that are not well understood, hurricanes go through "rapid intensification." As the name implies, the storm gets a lot stronger quickly. The technical definition is that the hurricane's winds increase 35 mph in twenty-four hours. On occasion, however, intensification is much more dramatic than that.

Hurricane Wilma set the record for a storm in the Atlantic Basin when it intensified from a 70 mph tropical storm to a 175 mph category 5 hurricane in twenty-four hours. Stunning! Luckily it was in the open waters of the western Caribbean at the time.

The far more dangerous scenario happened in 1935, 1957, and 2004, and no doubt other years as well. The Labor Day Hurricane in the Florida Keys, Hurricane Audrey, and Hurricane Charley all intensified rapidly in the twenty-four hours before hitting land.

Notable Rapid-Intensification Hurricanes

Year	Storm name	Wind decrease
1935	Labor Day	85–150 mph*
1957	Audrey	90–145 mph
2004	Charley	105–145 mph**

*From Sunday afternoon to Monday afternoon (Labor Day). Landfall came Monday evening with sustained winds of 160 mph or more.

**More important, Charley's maximum winds went from 110 to 145 mph in the six hours before landfall.

In 2006, Super Typhoon Cimaron went from a 40 mph tropical storm to a 160 mph–180 mph (satellite estimates varied, there were no direct measurements), category 5 hurricane in forty-eight hours. At the same time the pressure was estimated to have dropped an incredible 118 mb. Cimaron hit the Philippines with maximum sustained winds of at least 160 mph. This super-typhoon was the second of an incredible one, two, three, four punch to hit the Philippine island of Luzon in October and November of 2006.

Super Typhoon Cimaron approaches the island of Luzon in the Philippines on October 29, 2006.

The world record for intensification appears to be held by Super Typhoon Forrest when it was about 600 miles west-northwest of Guam. Beginning on the evening of September 22, 1983, Forrest's central pressure dropped 101 mb in twenty-four hours! In forty-eight hours the pressure dropped 120 mb, just slightly more than Cimaron's. But, remember, these days central pressure is estimated from satellite images. Forrest's pressure was measured by aircraft reconaissance, a more precise system, so the comparison isn't perfect.

All of the Atlantic storms had one thing in common: a small eye. Rapid intensification usually coincides with the shrinking of the eye, which makes sense according to the physical principle of conservation of angular momentum. But a small eye is not required. Hurricane Katrina's eye shrank from about 50 miles across to 30 miles or less (still fairly large), while the maximum winds when from 115 mph to 175 mph.

Rapid intensification of a hurricane near the coast is the greatest concern of forecasters. In spite of admonitions to be ready, people simply will not evacuate and otherwise prepare for a category 2 hurricane the way they will for a category 4 or 5. On occasion, however, storms rapidly intensify, and bad things—and potentially catastrophic things—happen.

Estimating Wind Speed

Getting air moving. A vacuum cleaner creates suction by creating low pressure inside the machine. The outside, higher-pressure air rushes into the hose toward the low pressure. When the system is turned around and you use the vacuum to blow air, the machine creates high pressure so the air rushes out toward the lower pressure of the room.

The phenomenon we call wind works the same way. The air around us is moving from areas of higher pressure toward areas of lower pressure. The lower the pressure, the faster the wind blows. (It's like having a stronger vacuum cleaner.)

Measuring the wind is difficult. Most of the time it's impossible to measure the highest winds in a hurricane. When the hurricane is over water, obviously, nobody is at the surface of the ocean with an anemometer. But, even when it's near land, there usually isn't a wind instrument at the exact place where the highest winds are.

Modern instruments such as Doppler radar and the new step-frequency microwave radiometer on reconnaissance aircraft are making the estimates better, when they are available, but they still don't give us a total picture of the strongest winds at ground level. Remember, when we say, "It's a 100 mph hurricane," we mean that the maximum sustained winds anywhere in the storm are *estimated* to be 100 mph. That's why methods for estimating the sustained winds are important.

Measuring the atmospheric pressure. On the other hand, atmospheric pressure in a hurricane is relatively easy to measure. Reconnaissance aircraft drop instrument packages called dropwindsondes into the center of the storm. They radio temperature, wind, pressure, and their location back to the plane as they fall through the atmosphere. In a well-developed storm, if the wind where the instrument falls is calm, it's assumed that the pressure measurement at that location is at or near the point with the minimum pressure. Using onboard radar and other instruments, these "drops" have gotten pretty accurate.

The pressure/wind formula. Because they are related, you can do a calculation to estimate the wind if you know the pressure. Researchers at the National Hurricane Center recently analyzed past hurricanes and came up with this formula:*

$$\text{Maximum Winds (mph)} = 9.615 \times (1015.8 - \text{Pressure (mb)})^{-.6143}$$

Okay, I know that's a bit obtuse if you haven't studied some math. You need a scientific calculator to put in the .6143 exponent. A simpler formula is:

---

*Formula from a recent paper by Dan Brown, James Franklin, and Chris Landsea at the National Hurricane Center.

$$\text{Maximum Winds (mph)} = 1375 - (1.315 \times \text{Pressure (mb)})$$

The second equation is not as accurate, especially for storms that are at the strong or weak end of the scale, but it will give you a number within 5 percent or so of the other formula. For strong storms with extremely low pressure, the second formula gives maximum-wind numbers that are a bit too high.

All storms aren't the same. Use the equations with caution. They only give an estimate of the maximum wind in an *average* hurricane with a given minimum pressure. The strength of high-pressure systems nearby, the size of the eye, and other factors can throw off the relationship significantly.

The Atlantic Hurricane Reanalysis Project. A major project is under way to reexamine the hurricane record back to 1851, and apply modern knowledge to the historical information that exists. As of this writing the storm tracks and intensities have been scrutinized through the 1920s, but not all of the updated information has gotten the final stamp of approval from the review committee.

The wind/pressure relationship is important in determining what the maximum winds might have been in a storm for which the records are spotty. To increase the accuracy of the estimated wind speed, modern records were analyzed to determine how the wind speeds and central pressures relate to each other *in various parts of the Atlantic Basin.*

It was found that, for a given pressure, wind speeds in the Gulf of Mexico are slightly lower than those in the Atlantic for weak storms and slightly higher for strong storms. On the other hand, storms with a low central pressure have considerably lower maximum winds as they move north in the Atlantic because the cooler water limits the development of tall thunderstorm cells that can bring the strong winds aloft down to the surface.

The point is, many, many variables are involved that make the wind/pressure relationship extremely complex. And since wind speeds are difficult to measure directly—and were even more so before the late 1990s—the hurricane record can't be taken as gospel, just best estimates of what happened.

Hurricane Definitions and Terms

anemometer   A weather instrument that measures wind speed.

anticyclone   A high pressure system.

anticyclonic   Rotation like a high pressure system (also known as an anticyclone). Clockwise in the northern hemisphere.

Atlantic Basin   The Atlantic Ocean, the Caribbean Sea, and the Gulf of Mexico.

Beaufort wind scale   A method of describing wind speeds designed by British Admiral Francis Beaufort in 1831. See page 58.

center or circulation center   The location of minimum wind or air pressure at the surface.

center fix   The location of the center as determined by reconnaissance aircraft, satellite, radar, or ground observations.

center relocated   On occasion, developing systems form a new center. In an advisory, the center is said to be *relocated* to the new location as opposed to having moved there. Sometimes centers are relocated because they were not well defined on nighttime satellite images, but become visible in daylight.

central dense overcast, or CDO   The region of solid cloudiness near the core of a tropical storm or hurricane.

central pressure   The minimum pressure in a tropical storm or hurricane.

concentric eyewall   A well-developed hurricane often goes through cycles in which the eyewall diameter shrinks and another eyewall, larger in diameter, forms. While both rings of strong cells exist, they are called concentric eyewalls.

convection   Technically, it's anything in atmosphere that moves in a vertical direction. The word is normally used with reference to a storm cell, however, in which the cloud tops are cold; that is, the clouds extend high in the atmosphere. Strong convective cells normally have heavy rain and are often thunderstorms.

coriolis force   The tendency of winds in the northern hemisphere to want to turn to the right when they blow over a long distance due to the rotation of the earth. The spinning of water in a toilet or going down a drain is *not* related to the Coriolis effect, the distances involved are too small. Named for Gaspard-Gustave Coriolis, a nineteenth-century French scientist.

cyclone The generic name for any low pressure area with a closed circulation. Also, hurricanes are called "tropical cyclones" in the Indian Ocean and near Australia.

cyclonic Having a rotation like a low-pressure system (cyclone). Counterclockwise in the northern hemisphere.

deepening The central pressure is dropping.

dropsonde or dropwindsonde An instrument package dropped from an airplane to measure the atmosphere. As the package parachutes to the surface of the ocean, it radios temperature, humidity, pressure, and so on back to the aircraft. A GPS radio sends back exact positions. With this information, the speed of movement can be calculated, yielding the speed of the winds in the storm.

Dvorak technique A method of estimating the intensity of a tropical system from satellite images, first proposed by Vernon Dvorak in 1973. Forecasters still use the system to determine the estimated wind speed of a storm far from land by comparing the cloud pattern as seen from the satellite to a chart matching patterns and wind speeds.

eye The roughly circular area of comparatively light winds in the center of a hurricane. Winds may be calm near the center of the eye. The strongest winds in a well-developed hurricane normally occur just outside the eye.

eyewall An organized ring of strong cells that surround the eye. Strong hurricanes often go through cycles in which the eyewall shrinks in diameter and another, larger one forms. This is called the "eyewall replacement cycle."

extratropical When a storm has become extratropical, it means that warm ocean water is no longer providing the energy. Extratropical storms in northern latitude get their energy from the interaction of warm and cold air masses. Storms normally become extratropical when they travel north and interact with the northern jet stream.

eyewall replacement cycle, or concentric eyewall cycle Intense hurricanes often have a relatively small eye, perhaps five to fifteen miles wide. Over time, rainbands can form an outer ring. Energy and moisture feeding into the storm is stopped by this ring from reaching the eyewall, so the inner structure dies, and the outer ring becomes the new eyewall. During this process the storm weakens. If the new eye contracts, the storm can regain its strength.

feeder band   Spiral bands of clouds and strong winds that curve into the center of a hurricane. Between feeder bands the weather can be quite good.

filling   The central pressure is increasing.

Fujiwara effect   When two storms of similar intensity come close enough together that their circulations interact, they tend to rotate around each other in a counterclockwise direction. Dr. Sakuhei Fujiwara, a Japanese scientist, described the phenomenon in water vortices in 1921. On average, the storms have to be within about 900 miles of each other to do the Fujiwara.

gale   An extratropical storm that is producing wind between 32 and 54 mph.

gale warning   Similar to a low-end tropical storm warning for a low-pressure system that is *not* a tropical storm or hurricane (i.e., an extratropical storm). Gale-force winds are between 32 and 54 mph.

global models   Computer weather forecasting programs that produce projections for the entire globe (or sometimes one hemisphere). Other models focus only on the area around the storm.

hurricane   A tropical cyclone with sustained surface winds higher than 73 mph in the Atlantic Basin and the North Pacific (east of the International Dateline.) If there are believed to be hurricane-force winds anywhere in the circulation, the National Hurricane Center will designate the system a hurricane. Often, in weak storms, the only hurricane-force winds occur offshore, decreasing to tropical-storm force winds over land.

hurricane season   The time of year when hurricanes most often occur. By definition, June 1 to November 30 in the Atlantic Basin, and May 15 to November 30 in the Eastern Pacific. However, tropical storms or hurricanes have formed in every month of the year.

hurricane warning   An alert that sustained winds of at least 74 mph are *expected* in the specified area within about twenty-four hours or less. Preparations should be rushed to completion.

hurricane watch An alert that hurricane conditions are *possible* in the specified area within about thirty-six hours or less. The first stages of preparation should begin for many people.

IR (infrared) satellite images The infrared sensor on the satellite measures heat. Since clouds are generally colder than the earth, they can be "seen" by this sensor twenty-four hours a day. Stronger cells extend higher in the atmosphere and are therefore colder at the top. A computer often adds artificial color to the various temperatures to make these cells stand out.

landfall The time when the *center* of the tropical cyclone crosses the coastline, *not* when the strong winds first arrive.

loop current An area of *deep* warm water. Hurricanes stir up the water as they move and often bring to the surface cold water that can cause the storm to weaken. When a storm tracks over a loop current, however, there is no cold water to bring up. Hurricanes often strengthen over loop currents in the Gulf of Mexico.

major hurricane A category 3, 4, or 5 hurricane on the Saffir-Simpson scale. Major hurricanes are responsible for about 80 percent of the damage.

Madden-Julian Oscillation (MJO) A 30- to 60-day cycle in tropical weather activity. Discovered by scientists Roland Madden and Paul Julian in 1971 while examining western Pacific weather records. The intensity of thunderstorms in the tropics increases and decreases in the roughly defined cycles.

QuikSCAT Short for "quick scatterometer." A special type of satellite that uses a radar beam aimed at the ocean to estimate the winds near the surface. The readings can be contaminated by rain, but the images are useful when reconnaissance-aircraft data aren't available. The satellite orbits around the poles and makes swaths about 1,100 miles wide as it passes over the tropical oceans. Any one spot normally gets covered twice a day.

radius of maximum winds The distance from the center of the storm to the band of strongest winds. In a well-developed hurricane, the strongest winds normally are immediately outside the eye. In a weakening or disorganized storm, however, the strongest winds may be a considerable distance from the center.

rapid intensification   When an Atlantic storm strengthens by 35 mph in twenty-four hours.

rawindsonde   An instrument package attached to a weather balloon to measure temperature, humidity, and pressure. The balloon's movement is tracked by a radio system to determine its movement, and therefore the speed and direction of the wind at various levels in the atmosphere.

reduction factor   The amount of difference between the winds at flight level, normally 10,000 feet, and at the surface. This number is now known to vary with the strength of the storm, but the average surface wind is 80 to 90 percent of the speed measured at flight level.

Saffir-Simpson Hurricane Scale   A 1 to 5 rating system for hurricanes based on their potential damage and storm surge. See page 148.

step (or "stepped") frequency microwave radiometer (SFMR)   An instrument aboard some reconnaissance aircraft to remotely measure surface wind speeds by analyzing signals received from the sea spray. The more spray, the stronger the wind.

storm surge   The increase in the water level due to a hurricane, not counting any contribution from the astronomical tides or waves. See page 117.

storm tide   The increase in water level from storm surge plus the contribution of astronomical tides, but not counting the height of waves.

storm warning   Beware of this term. It's confusingly similar to "tropical storm warning." An alert for winds over 55 mph due to a low-pressure system that is *not* a tropical storm or hurricane (i.e., an extratropical storm). The warning system for *non*tropical systems is comprised of *gale warning* and *storm warning*. Generally, these warnings are only of interest to mariners.

subtropical depression   A storm system not associated with a front that has characteristics of both tropical and extratropical cyclones with maximum winds less than 39 mph. Normally these systems are asymmetrical in shape and have a wide zone of maximum winds located well away from the center. On occasion these systems evolve into tropical storms and even hurricanes. See page 107.

subtropical storm   A storm system not associated with a front that has characteristics of both tropical and extratropical cyclones with maximum winds over 39 mph. Normally these systems are asymmetrical in shape and have a wide zone of maximum winds located well away from the center. On occasion these systems evolve into tropical storms and even hurricanes. See page 107.

supertyphoon   A typhoon with winds over 150 mph. Equivalent to a strong category 4 or a category 5 hurricane.

synoptic times   The four times around the clock when data is collected worldwide: 0000Z, 0600Z, 1200Z, and 1800Z, which are the same as 8:00 P.M., 2:00 A.M., 8:00 A.M., and 2:00 P.M. EDT.

trade winds   The steady wind flow from the east and northeast across the region between 30°N and the equator. The most consistent part of the atmospheric circulation pattern. The winds are named for the trading ships which used them to travel from Europe to the New World. Tropical waves travel west in this flow.

TRMM satellite   A satellite equipped with a radar system for measuring and studying tropical rainfall. TRMM stands for "tropical rainfall measuring mission."

tropical depression   An organized tropical system with a defined circulation and maximum sustained winds of less than 39 mph.

tropical disturbance   A discrete weather system of thunderstorms in the tropics not associated with a frontal system that maintains its characteristic for twenty-four hours or more. Tropical disturbances are generally the "seeds" from which tropical storms and hurricanes grow. Often incorrectly used interchangeably with *tropical wave*.

tropical storm   An organized tropical system with a defined circulation and maximum sustained winds between 39 and 73 mph.

tropical storm watch   An alert that tropical storm conditions are *possible* in the specified area within about thirty-six hours or less.

tropical storm warning   An alert that sustained winds of at least 39 mph are *expected* in the specified area within twenty-four hours or less.

tropical wave   Also called an easterly wave. A defined kink or bend in the trade winds blowing from east to west over the tropical Atlantic. On a map that shows the direction of the wind, it appears as a hump or a wave in the overall wind flow. Often incorrectly used interchangeably with *tropical disturbance*.

TUTT   Tropical upper tropospheric trough. An upper-level low-pressure area in the tropics. Sometimes TUTTs can create an upper wind flow that may inhibit the strengthening of tropical systems. At other times, however, if a TUTT is just far enough away from a tropical storm or hurricane, it can aid the strengthening process.

visible satellite images   Visible satellite pictures are like a photographs from space. They have four times the resolution of IR images, but are only available during daylight hours.

water vapor satellite images   A sensor on the satellite detects the level of water vapor in the middle and upper layers of the atmosphere. These are often useful for seeing steering currents or areas of dry air that might affect a storm's intensity.

wind shear   The wind may blow at a different speed or in a different direction at various levels in the atmosphere, a phenomenon known as wind shear. Tropical cyclones can only strengthen if the wind is fairly uniform with increasing height, allowing the eye-structure to develop; that is, there is light wind shear.

Z-time   The same as Greenwich mean time. Four hours earlier than eastern daylight time. Five hours earlier than eastern standard time. 1200Z = 8:00 A.M. EDT, for example.

4

National Hurricane Center and National Weather Service Bulletins

Information for residents in a zone that might be affected by a hurricane is issued by both the National Hurricane Center in Miami and the local office of the National Weather Service having responsibility for that particular zone. Unfortunately there is no one-stop shop for all of the critical, ever-changing information on a storm, its consequences, and the recommended preparedness actions.

The Hurricane Local Statement, see below, is as close as there is to a comprehensive bulletin, but it is not a complete view of the storm and has other faults. See "Hurricane Bulletin Tips," page 135. So, for now, to get the full picture, you have to digest both the advisory package from National Hurricane Center and the bulletins from your local National Weather Service office.

NHC Advisory Package

To fully understand the National Hurricane Center's forecast, you have to look at the *entire package* of bulletins and graphics that is issued four times a day. (No, this is not the most efficient imaginable system).

The full set of bulletins and graphics that are issued in each package are:

- **Public Advisory.** Storm information in plain English. Page 155.

- **Forecast Advisory** (formerly, and still sometimes, called the Marine Advisory). Somewhat technical current and forecast information in a combination text and tabular format. Page 158.

- **Forecast Discussion.** A technical explanation of the rationale used by the forecaster in preparing the advisory. Page 163.

- **Probabilistic Surface Wind Speed Text Product** (*New* for 2006). An extensive table of numbers giving the odds of certain strength winds occurring at various locations. Page 164.

- **Watch/Warn 3-Day and 5-Day** graphics (the forecast cones). The "official" cones, similar to those shown on television news programs. Page 164.

- **Wind Speed Forecast and Probabilities** graphic. A different kind of "cone" that shows the areas most likely to receive tropical storm force winds. Page 167.

- **Wind Speed Probability Table.** A table showing the odds that the maximum wind speed in the storm will be different than the forecast. Page 168.

- **Cumulative Wind Distribution** graphic. A swath of the areas that have been affected by tropical-storm and hurricane-strength winds. It is not related to the forecast.

Intermediate Public Advisories

The issuance schedule for public Advisories increases to every three hours whenever watches and warnings have been issued, then to every two hours when a hurricane gets within radar range of the U.S. coast, where the eye can be tracked minute by minute.

Bulletins from Local National Weather Service Offices

In addition to the bulletins and graphics above, when the storm is threatening the U.S. coastline, local National Weather Service offices issue important text bulletins associated with the storm.

NOTE: The word *local* here is a misnomer. "Local" NWS offices cover a region of many, often diverse, counties. A map of local NWS offices is available at www.srh.weather.gov.

- **Hurricane Local Statement.** A long—sometimes very long—text product with specifics on how the storm is expected to impact the local area, evacuation orders, and necessary precautions.
- **Inland Watches and Warnings.** Inland hurricane watches, inland hurricane warnings, inland tropical storm watches, and inland tropical storm warnings are issued to alert populations away from the coast of the effects of a hurricane. They are the equivalent of the similarly named alerts for coastal areas.
- **Tornado Warning.** Issued to warn for tornadoes and now also issued to convey the threat from extremely high winds near the eyewall of a strong hurricane (a confusing use of the tornado warning, in my opinion).
- **Flood Watch.** Issued in advance of storms that *might* bring flooding rain.
- **Flood Warning.** Issued when significant flooding is imminent.

Each bulletin or graphic conveys a piece of the forecast puzzle. The NHC advisories, in general, make specific statements about the current state and the forecast for the storm for the next five days, if the storm is expected to survive that long. The Probabilistic Surface Wind Speed Text Product and all the NHC forecast graphics listed above are designed to communicate the uncertainty associated with that forecast. The products issued by the National Hurricane Center deal with the storm itself and its broad-brush effects. The local NWS products give more details for specific cities and counties.

The tremendous volume of information contained in these advisories and bulletins, issued a least four times a day, is necessary because of one inescapable and critical fact: *Every hurricane forecast is uncertain.*

If National Hurricane Center forecasters knew exactly where the storm was going to go and precisely what it was going to do, one hurricane bulletin would cover it. But there is uncertainty in every aspect of the forecast. And some parts are more uncertain than others. That's why the hurricane advisory package is so long and complicated.

This forecast uncertainty is communicated in a variety of ways.

- **Forecast or wind-speed probabilities** used to generate text and graphics products are based on past errors in National Hurricane Center forecasts. The average errors over the past ten years are used to calculate the width of the cone or the likelihood that winds of a certain strength will occur. These uncertainties are *not* related to anything about the particular storm being forecast.
- The **subjective uncertainty,** including the confidence of the forecaster, is sometimes communicated in the Forecast Discussion.
- The **intrinsic uncertainty** in the forecasts is expected to be understood. The words and the system the NHC uses to communicate hurricane forecasts are precise when precision is not a reality. The assumption is that the user understands the built-in uncertainty. See page 170 for more on forecast uncertainty.

Remember, to fully understand the advisories from the National Hurricane Center, you must use and understand the entire package of information. Don't focus on any one part of the forecast package.

Other Bulletins with Tropical Cyclone Information

Tropical Weather Outlook. Issued four times a day during hurricane season by the National Hurricane Center listing any hurricanes, tropical storms, or depressions ongoing in the Atlantic Basin. Also, tropical disturbances that appear to have potential for development are indicated.

Tropical Weather Discussion. A technical description of the weather systems affecting the Gulf of Mexico, Caribbean, and the tropical Atlantic. Issued four times a day by the National Hurricane Center.

Tropical Cyclone Update. Issued by the National Hurricane Center when sudden changes have occurred in the storm or new information has come in between advisory times. When Hurricane Hunters measure significant changes in pressure or wind requiring adjustments in the next advisory, the Tropical Cyclone Update is a short bulletin indicating that important changes are under way.

Tropical Cyclone Position Estimate. When a hurricane is near the coast, the National Hurricane Center issues a short update of the storm center's position every hour that no advisories are being issued.

Special Tropical Disturbance Statement. On occasion a tropical disturbance quickly develops. The National Hurricane Center issues this bulletin as an alert that important changes are under way.

Hurricane Bulletin Tips

Movement message confusion. When a storm is moving errati-cally, the movement is better conveyed in the Public Advisory than in the Forecast Advisory. Words can more accurately describe what's hap-pening than numbers in a tablelike format. On these occasions, move-ment information plotted by computer programs that rely solely on the Forecast Advisory is often incorrect. The problem only occurs when a storm is moving slowly, so usually no one is endangered by the misin-formation. It does cause some confusion, however.

Hurricane Local Statement (HLS) overcomplexity. This bulletin is unnecessarily long and complicated. Only one HLS is issued by each NWS Forecast Office, regardless of the number of diverse locales in their service areas. As a result, users interested in the storm's expected effects in one city have to sort through confusing, irrelevant informa-tion that doesn't apply to them.

For example, the cities of Baltimore and Washington get the same Hurricane Local Statement despite their hurricane-impact issues being very different. This contributed to the media and the general public in the Baltimore area being surprised by the massive flood that came with Hurricane Isabel in 2003. The National Weather Service wasn't sur-prised, but the information for Baltimore was poorly communicated and mixed in with information for Washington.

In South Florida, the southeast coast from Miami to Palm Beach gets the same Hurricane Local Statement as the lower Gulf Coast. The two areas couldn't be more different. The HLS for Miami gets bogged down in irrelevant information for a mostly unpopulated, unrelated area.

You have to read these bulletins very carefully to determine what's important to you.

Inland hurricane watches and warnings. The question begs to be asked, "What is the difference between an inland hurricane warning and a hurricane warning?" The answer for most people is "Nothing." Within the National Weather Service, the NHC issues the hurricane warnings while the local NWS offices issue the inland hurricane warnings. But from the standpoint of the people who are threatened, the two bulletins mean the same thing: Prepare for a hurricane. The NHC/NWS draws a line, sometimes coincident with a county line, sometimes not, where the warning changes name.

For the general public, it's a distinction without a difference, and it's a cause of confusion. The National Weather Service would do well to look at this from the public's perspective.

Hurricane Names

Storms are named so we can keep them straight. Imagine if advisories were being issued for several unnamed storms at the same time. Confusion would reign. In addition, of course, the names give hurricanes personality. Imagine if Katrina were called simply the Great New Orleans Levee-Break Storm of 2005. It would be cumbersome.

2007 Names for the Atlantic, Caribbean, and the Gulf of Mexico

Andrea	Humberto	Olga
Barry	Ingrid	Pablo
Chantal	Jerry	Rebekah
Dean	Karen	Sebastien
Erin	Lorenzo	Tanya
Felix	Melissa	Van
Gabrielle	Noel	Wendy

Clement Wragge

Tropical systems were first named after people by Clement Lindsay Wragge, the colorful and controversial meteorologist in charge of the Brisbane, Australia, government weather office at the end of the nineteenth century. He initially named the storms after mythological figures, but later named them after politicians he didn't like. He fought with the government over the right to issue national forecasts. He lost and was fired in 1902.

Naming History and Rules

At one time, especially damaging storms were named simply to commemorate and remember the event. But today the World Meteorological Organization (WMO) is in charge of the names and the naming system.

Saints' names. The practice of naming damaging hurricanes goes back as long as anybody can remember. But the system for naming them was haphazard, not really a system at all. Strong hurricanes hitting the Spanish islands of the Caribbean got Catholic saints' names. The Santa Ana (in English, Saint Anne) Hurricane hit Puerto Rico on July 26, 1825, the date of the feast in honor of the Mother of the Blessed Virgin. San Filipe (third-century Saint Philip) got two hurricanes named after him because they both hit Puerto Rico on his day, September 13, one in 1876 and then a monster in 1928 that moved on to become the Great Okeechobee Hurricane. See page 49.

Where, when, and what name. Other storms were given names related to *where* they hit: the Galveston Hurricane (1900), the Great Miami Hurricane (1926), the Okeechobee Hurricane (1928); *when* they hit, like the famous Labor Day Hurricane (1935) that hit the Florida Keys; or *what* they hit, like the Rising Sun Hurricane that hit Charleston, South Carolina, on September 14, 1700. The *Rising Sun,* a Scottish ship, had just made it to port ahead of the storm, but the settlers making their way to the New World had not yet disembarked when the ship was smashed. All ninety-seven people aboard were lost.

Fictional Storm Called Maria, 1941

George Rippey Stewart wrote a bestselling novel in 1941 called *Storm*. Set in the San Francisco Weather Bureau office, it centers on a junior meteorologist who is tracking Pacific storms. He unofficially gives the storms women's names because, he says, they each have a unique personality. The focus of the book is a storm named Maria, but pronounced "Ma-RYE-a." Yes, the song in the Broadway show *Paint Your Wagon* named "They Call the Wind Maria" was inspired by this fictional storm. I can't say whether George Stewart was indeed the inspiration for the trend toward naming hurricanes that came along later in the decade, but it seems likely. In any case, he certainly had it right in describing hurricanes as having human characteristics. "Each one is different. There are the big bluffers, and the sneaks, and the honest dependable ones. Some will sulk for days and some will stab you in the back, and some walk out on you between night and morning, and some do exactly what you expect of them."

Great Atlantic Hurricane of 1944. A hurricane with many firsts. The first hurricane to be measured by a scheduled reconnaissance flight; the first hurricane to be viewed on radar (aboard the plane); and the first hurricane "officially" named by the Miami Hurricane Warning Office, which opened the previous year. See page 54.

World War II. Some accounts of typhoons that damaged ships of the Pacific Fleet during World War II refer to the storms by name. On December 18, 1944, a typhoon the military meteorologists named Cobra scattered Admiral William Halsey's fleet all over the Philippine Sea. On October 9, 1945, a storm named Typhoon Louise by the Guam Weather Bureau hit Okinawa, Japan, devastating the fleet of 300-some U.S. Navy ships that had sought safe harbor in Bruckner Bay. Ships crashed together, sank, and ran aground, delaying the postwar return of soldiers to the United States that had been scheduled to begin that day.

Hurricane George?—1947. You'll find references in the historical literature to the Miami Weather Bureau (see page "Milestones in Hurricane Science," page 95, naming the major hurricane that hit north of Fort Lauderdale on September 17, 1947, Hurricane George. I've looked through the advisories issued for that storm and don't see any reference to *George*.

Harry's Hurricane and Hurricane Bess, 1949. President Harry Truman was in Miami addressing the Veterans of Foreign Wars on August 25, 1949, at the same time the first hurricane of that season was moving through the Atlantic east of the Bahamas. The *Miami Herald* used "Harry's Hurricane" in a headline of a story about the storm. The second storm of that year was a major hurricane that hit the east coast of Florida. In the records at the NHC (but not the advisories) *Bess,* after Bess Truman, the first lady, is printed next to information about this storm. I cannot find it in the press, however, in spite of reports of that in various articles.

Hurricane Fox, 1950. The first storm to be named (using a naming scheme) in an official Weather Bureau advisory was Hurricane Fox on September 10, 1950. The San Juan office's advisory on that date was the first to include a storm's name, and all advisories after that had the name in the bulletin header. Hurricane bulletins at that time were being issued by the San Juan, Miami, Washington, and New Orleans Weather Bureau offices.

Military alphabet, 1950, 1951, and 1952. *Fox* came from the World War II phonetic alphabet, which was adopted as the first official hurricane-naming scheme (likely because most of the people forecasting the weather at that time had come out of the military). The names were Able, Baker, Charley, and so forth and included such oddities as Tropical Storm How and Hurricane Jig. The same system continued from 1950 to 1952, so each of those years also had storms named Able though Fox. The year 1952 only had six named storms. The most significant storm of that era was Hurricane King, a category 3 storm that hit Miami and Fort Lauderdale on October 17, 1950.

NOTE: I can't find any record of why the naming didn't start in 1950 until Hurricane Fox. Suddenly the name appears on the advisories. In the 1950 season summary, the narrative describing the storms makes no reference to any names at all, so the naming process was obviously not well established. The listing of the year's storms begins with Able, however. If you have any information, please let me know at www .hurricanealmanac.com.

Women's names begin, 1953. A list of twenty-three women's names was adopted in 1953, beginning with Alice and ending with Wallis (not the most common of names, but the Duchess of Windsor, the former Wallis Simpson, was a very prominent personality at the time). Orpha and Una were also on the list. (Anybody know an Orpha or an Una?) The plan was to use the same list every year, as had been done previously with the military alphabet.

The year with two Alices, 1954. The same list was used in 1954, except that Gilda was used instead of Gail. (I assume that was because *Gail* sounds just like *gale*, which is the technical name for a windstorm over the water with winds between 32 and 63 miles per hour. With Gilda there could be no confusion.) But talk about confusion. If you look at a list of 1954 storms, you'll find two storms named Alice, one in June and one in December. How could that be? (See 1955, below.) There were also three major hurricane events on the East Coast in 1954, Hurricanes Carol, Edna, and Hazel. See page 59.

A new list of names, 1955. On January 1, 1955, a strong storm was noted northeast of the Lesser Antilles. Since the Weather Bureau was still working under the scheme of the previous two years, it was designated Hurricane Alice. The system had been noted previously, but only as a "tropical disturbance." Subsequent examination of the available data has determined that the storm actually formed before the New Year, so in the official record it is the second Alice of 1954. See page 74.

The two powerful hurricanes of 1954—Carol, which hit Long Island and New England, and Hazel, which did significant damage from North Carolina to Toronto—caused the naming scheme to be changed for 1955. To avoid confusion with those famous storms, a new list was devised beginning with Brenda (remember, Alice had already been used in January).

New list each year, 1956–59. New lists were developed with completely different women's names for each of these four years.

Rotating lists begin, 1960. Four permanent lists were developed for the beginning of the 1960s, using the late-fifties lists as a starting point. The idea was to reuse the first list in the fifth year, with replacements for any high-profile storms. (The 1962 list looks a lot like 1958, and 1963 like 1959, but there are some differences.)

More lists added, 1971. The number of lists of women's names was expanded to ten in 1971, but otherwise the system was left unchanged.

Men's names added and lists cut, 1979. In keeping with the changing sensibilities in society, the hurricane-name lists included men's names for the first time in 1979. The first male storm was Hurricane Bob, which hit near New Orleans in July. The second was category 5 Hurricane David, which killed over 2,000 people in the Dominican Republic on September 1. The third one, Frederic, devastated southern Alabama later the same month. So it was an auspicious start for the men's names. In addition, the number of lists was cut to six, the system we use today. The 1979 list was used again in 1985, except David and Frederic were retired and replaced by Danny and Fabian.

Adding men's names didn't go down easily with some of the more conservative countries on the World Meteorological Organization (WMO) committee. As a compromise, Spanish and French names were included in the list. So, the new names were not only gender-diverse, but culturally diverse as well.

Retired hurricane names. The criterion for retiring a hurricane's name is subjective. The National Hurricane Center says that a name should be retired if "a storm is so deadly or costly that the future use of its name on a different storm would be inappropriate for obvious reasons of sensitivity." In practice, the names are retired when they do significant damage or cause a significant loss of life. At the yearly regional meeting of the WMO, the previous season's storms are evaluated and the decisions are made on which names get retired and what their replacements will be. Any member of the committee can request that a name be retired. For a complete list of retired names see www .hurricanealmanac.com.

A Retirement Record

Dennis, Katrina, Rita, Stan, and Wilma were retired after the 2005 season, the most names ever retired in one year. The previous record was four, in 1955, 1995, and 2004. Prior to the 2005 season, sixty-two hurricane names had been retired.

Naming exceptions. Before the lists finally became standardized in 1979, there were some exceptions to the rules. Carol was reused during the 1965 season, but then retired in honor of the 1954 storm. (Originally the retirement period was set at ten years.) Similarly, an inconsequential Tropical Storm Edna formed in 1968, but the name was retired in honor of Hurricane Edna that hit North Carolina and Maine during the extremely busy hurricane summer of 1954. Other names disappeared instead of being retired, like the 1963 names, Cindy, Debra, Ginny, and Helena. In 1967 they were replaced for no obvious reason by Chloe, Doria, Ginger, and Heidi.

Unnamed hurricanes and tropical storms. Many storms are listed, even in the years after 1952, with no names. Most often that's because these systems were not known to meet the criteria necessary to be designated a tropical or subtropical cyclone until after the storm had died out, sometimes even years later. A reanalysis of the radar and satellite data or late-arriving wind or pressure observations from ships or other credible observers might be enough to retroactively designate a system for a given year's list. Some systems were added to the 1960s lists when the early satellite pictures were reexamined, but most often it was because of late-arriving data. Two tropical storms, a subtropical storm, and two hurricanes were retroactively added to the 1969 list, making this already busy season even busier. There were twelve hurricanes that year, which tied the record. The mega season of 2005 finally broke that record with fifteen.

Neutercanes. In 1972 the term *neutercanes* was used for storms that had some characteristics of a *tropical* system but *nontropical* aspects as well. We call them *subtropical* cyclones today, but that classification didn't exist at the time. The then-National Hurricane Center director, Robert Simpson, proposed the new classification for these storms, and it was put into use that year. The modern military phonetic alphabet (as opposed to the old World War II alphabet used in the early 1950s) gave the storms their names, so the first system was named Neutercane Alpha in May. Neutercane Bravo came along in August. Advisories were written for Neutercane Charlie in September, and Delta formed in November. Alpha was the only neutercane to hit land, coming ashore near Brunswick, Georgia, on May 27 with winds of about 45 mph. In 1973, Neutercane Alfa (they changed *Alpha* to *Alfa* . . . anybody know why?) moved north offshore of the mid-Atlantic and Northeast coast in late July and early August. That was the last time these names and the term *neutercane* were used. Today these storms are classified as *subtropical depressions* and *subtropical storms* (see page 128).

Subtropical storms. Today, subtropical storms are named using the same list as tropical storms and hurricanes. This makes sense because subtropical cyclones often take on tropical characteristics. Imagine how confusing it would be if the system got a new name just because it underwent internal changes.

There is no subtropical classification equivalent to a hurricane. The assumption is that once a storm got that strong it would have acquired tropical characteristics—and therefore be called a hurricane—or have merged with an extratropical system in the North Atlantic and lost its name altogether. On October 24, 1979, a subtropical storm briefly reached hurricane strength as it approached Newfoundland, Canada. It quickly combined with another low-pressure system. It was never named.

The November 1, 1991, view from the GOES satellite of the huge cloud swirl related to the large low pressure with a small hurricane at its center. This configuration is unusual, but not unheard of. Courtesy NASA.

The Perfect Storm. The most famous unnamed storm was Hurricane No. 8 in 1991. Now it is best known as the Perfect Storm. It started as a typical autumn cold front moving into the North Atlantic. At the same time, however, Hurricane Grace was drifting near Bermuda.

Often a low-pressure system (called a wave) will form a cold front over the ocean. But the tropical system nearby made this situation different. The wave combined with the moist tropical air from the remnants of Grace, so the low that developed southeast of Newfoundland on October 28 was unusually large and strong. With winds gusting to hurricane strength, it caused major coastal flooding in New England, heavy snow, and a number of fatalities, including the crew of the fishing boat *Andrea Gail*.

The system weakened as it moved southwest, but an inner core developed resembling a tropical storm on August 31. The next day, a Hurricane Hunter flight found hurricane force winds in the storm's small tropical core, but the National Hurricane Center decided *not* to name the system. The forecast called for the storm's center to stay offshore, and since the system had already caused so much damage, there was concern that naming it a hurricane would cause unwarranted anxiety in the public.

Naming Rules for Unusual Situations

What happens if all twenty-one names in the Atlantic list get used up? This, of course, happened in 2005, so the National Hurricane Center went to Plan B, the Greek alphabet. Storm number 22 was Tropical Storm Alpha, then came Beta, Gamma, and so on. A problem came to light, however. Suppose Alpha had turned into a destructive hurricane and the name was "retired." Then what? You can't delete a letter from the alphabet. The solution is to attach a year number to the Greek-letter name. So, the twenty-second storm of 2008 (if there is one) will be Alpha 2008.

It was proposed that a second standby list of names be designated, to be used only in hyperactive seasons. This way, if an overflow-list storm is noteworthy, retiring the name and replacing it with another one is straightforward.

In 2005 this problem almost occurred. On October 4 a subtropical storm formed about 190 miles southwest of the Azores. Winds near 50 mph affected the islands that evening. The system was not named by the National Hurricane Center; it was not clear at the time that the system had any tropical characteristics, and it was right after Hurricane Rita. *If* it had been named, it would have been Tammy, and all of the subsequent names would have been pushed down. As a result, Hurricane Wilma would have been Hurricane Alpha, and the question of retiring a letter of the alphabet would have been front and center.

Suppose a storm weakens and then restrengthens? Does it get a new name? No. As long as the circulation can still be identified as largely that of the original system, the regenerated storm will get the same name. This is subjective, however. In 2004, the system that was Hurricane Ivan split into two identifiable circulations as it was dying out near Washington, D.C. One, mostly in the midlevels of the atmosphere, moved northeast, combining with another low pressure system to create hurricane-force winds in Nova Scotia. The other swirl moved south and then west, eventually redeveloping into a tropical depression and, briefly, a tropical storm over the Gulf of Mexico. This system was once again given the name Ivan.

But, in 2005, Tropical Depression #10 lost its circulation on August 15. Then on August 23, a system just east of the Bahamas composed, at least partially, of the remnants of TD #10 was upgraded and named Tropical Depression #12. Satellite pictures showed that another weak disturbance was in the area as well and had been absorbed into the new system. The situation was ambiguous enough that the old name, Depression #10, was not reused. (TD #12 went on to grow into Hurricane Katrina.)

What happens if a storm moves from the Gulf of Mexico to the Pacific? Does it keep the same name? The answer to this question has changed. Now, the name stays the same if, and it's a big if, the system maintains its circulation. If it weakens into a tropical wave, so that it's no longer at least a tropical depression, it would be assigned a new, Pacific name if it regenerates. If, however, the National Hurricane Center issues continuous advisories on the system, the same name will carry through. The same rules apply to systems moving from the Pacific into the Gulf or Caribbean.

This situation rarely comes up because the mountainous terrain of Mexico and Central America most often disrupts the surface circulation so that it is no longer identifiable by the time the system reaches the Pacific.

The last time this happened was before the rule was changed in 2000. In 1996, Hurricane Cesar moved from the Caribbean across Central America and was renamed Tropical Storm Douglas in the Pacific. That storm grew into a Category 3 hurricane before dissipating.

In 1989, Hurricane Cosme came ashore on the Pacific coast of Mexico and moved north. The remnants redeveloped and this storm was named Allison. It dumped over two feet of rain at Winnfield, Louisiana, near Shreveport, and caused significant flooding. This Allison shouldn't be confused with 2001's Tropical Storm Allison, which caused massive flooding in Houston and the surrounding areas. The name Allison was retired after the 2001 storm.

Naming/numbering "invest" systems. Before a tropical system is sufficiently organized to be classified a tropical depression, but appears to have some potential to strengthen, it is given "invest" status. The National Hurricane Center gives each of the disturbances it's watching a number and a letter. The numbers run 90 through 99, and then back to 90. If the invest is in the Atlantic Basin, the number is followed by an *L*, in the eastern Pacific they use an *E,* and in the central Pacific south of Hawaii they use *C*. Other basins have other letters.

So, for example, Invest 95L would be a system in the Atlantic designated for investigation by computer models and possibly aircraft reconnaissance.

After a system is designated a depression, it gets a lower number. For example, the first depression of the season in the Atlantic gets the technical name, 01L. During the 2005 hurricane season there were thirty-one depressions and named storms, so the last one was 31L Zeta. To avoid confusion with upgraded systems, the disturbances are designated with numbers in the 90s. The 80s are reserved for tests or exercises.

You can see which systems are currently being "investigated," including their numerical names, at www.nrlmry.navy.mil/tc_pages/tc_home.html.

Different Naming Rules Elsewhere in the World

The rules vary for naming hurricanes-typhoons-cyclones around the world.

Eastern Pacific. In the Eastern Pacific Ocean (east of 140°W) the rules are substantially the same as in the Atlantic. Six lists of names rotate, and they start at the top of the list each year. The difference is that the yearly lists are three names longer. X, Y, and Z names are included. But they are tricky. The three names at the end of the alphabet alternate every *two* years, not every six. Only Xina and Xavier, York and Yolanda, and Zelda and Zeke are on the lists.

Central Pacific. The rules in the zone between 140°W and the International Dateline, roughly the area south of Hawaii, are somewhat different. There they use four lists of twelve names of Hawaiian decent, so they have a total of forty-eight names to work with. They do *not* start at the top of the next list each year, however. The names are used in order, so the first storm in a season can start with any letter, depending on where the previous year left off. This area only averages four or five storms a year, so they can get by with fewer names.

Western Pacific. The naming situation west of the International Dateline has become complicated. Fourteen countries and territories that are affected by typhoons in the region contributed ten names each. So there are 140 names to work with. In general, however, they are

Asian words from various languages for birds, flowers, and trees. There are a few men's and women's names mixed in.

The names contributed by the United States (because of the affected U.S. territories in the region) are Maria, Francisco, Higos, Omais, Roke, Utor, Chataan, Etau, Aere, and Vicete. Maria, Francisco, Roke, and Vicente are Chamorro names (the language of Guam and the northern Mariana Islands, which sounds a bit like Spanish). Utor and Aere are Marshallese (the language of the Marshall Islands) for "squall line" and "storm." Omais and Etau are Palauan (from the former U.S. territory of Palau, southwest of Guam) meaning "wandering around" and "storm cloud." Matmo means "heavy rain" on Guam.

The Philippines, however, uses a separate set of names for storms that come into their area. Systems get Filipino names when they reach depression status because even weak systems can be dangerous on the mountainous islands. Also, the authorities there feel that people will respond better to local words and understand better that the storm is a local threat.

In Japan, generally they don't use the international community's names in the press. The typhoons are numbered by the year. So T5415, which was officially named Typhoon Marie, was the fifteenth storm of 1954 in the Pacific Basin.

Saffir-Simpson Hurricane Scale

The Saffir-Simpson Hurricane Scale gives an estimate of the potential damage and flooding that can be expected with landfalling hurricane. The 1 to 5 scale works for two reasons:

1. Damage occurs in stages. When the force of the wind exceeds that ability of a structure (or anything exposed to the storm) to withstand it, damage begins. At another wind speed, other items fail. These thresholds account, to some degree, for the steps in the Saffir-Simpson scale.
2. The storm's maximum wind is only an estimate—the best guess of the National Hurricane Center to the nearest 5 mph or 5 kt based on the available data. Since the maximum value is never

precisely known, it makes sense to group hurricanes by the level of damage they do instead of concentrating on the specific wind-speed number.

NOTE: I have modified the descriptions of the effects of each Saffir-Simpson category storm from the "official" wording. In my opinion they all need to be completely reworked. When the effects from a given category storm were written, the coastal zones were not as highly developed as they are today, so the old text doesn't fit today's reality. For example, the category 2 portrayal of potential damage doesn't sound at all like Hurricane Wilma in South Florida. Wilma was much worse because it went through a densely populated metropolitan area with many expensive properties.

Note that the storm-surge numbers listed are only estimates. The actual values vary tremendously, depending on the nature of the coastal area where the storm comes ashore. Also, it is well-known that there is not a direct correlation between central pressure and wind speed. The pressure numbers are average values.

Category 1: 74–95 MPH—Minimal Damage

No major damage to well-built structures. Damage primarily to poorly built buildings, some high-rise windows, unanchored mobile homes, shrubbery, and trees. Some damage to poorly constructed signs. Evacuations may be ordered for areas immediately adjacent to the water. Storm surge generally 4 to 5 feet. Central pressure approximately 980 mb.

Category 1 hurricanes. Hurricane Katrina when it hit Miami-Dade County in 2005; Hurricane Lili when it hit southern Louisiana in 2002.

Category 2: 96–110 MPH—Moderate Damage

Some roof, door, and window damage to buildings. Considerable damage to exposed high-rise glass, shrubbery, and trees, with some trees blown down. Considerable damage to mobile homes, poorly built buildings, signs, and piers. Small craft in unprotected anchorages may break their moorings. Evacuations often ordered for areas near the water. Storm surge generally 6 to 8 feet. Central pressure 965–979 mb.

Category 2 hurricanes. Hurricane Wilma when it hit metropolitan South Florida in 2005; Hurricane Frances when it hit the central Florida coast in 2004.

Category 3–5 hurricanes are called *major hurricanes*.

Category 3: 111–130 MPH—Extensive Damage

Structural damage to some residences is likely, with some exterior wall and glass failures in large buildings. Damage to shrubbery and trees with foliage blown off. Large trees blown down. Mobile homes and signs are destroyed. Flooding near the coast destroys smaller structures, with larger structures damaged by battering waves and floating debris. Evacuations will likely be ordered for areas susceptible to storm-surge flooding. Storm surge generally 9 to 12 feet. Central pressure 945–964 mb.

Category 3 hurricanes. Hurricane Katrina when it hit Louisiana and Mississippi in 2005; Hurricane Ivan when it hit the Pensacola area in 2004.

Category 4: 131–155 MPH—Extreme Damage

Extensive exterior failures in large buildings with some complete roof failures on residences. Shrubs, trees, and all signs are blown down. Complete destruction of mobile homes and poorly built buildings. Extensive damage to doors and windows. Major damage to lower floors of structures near the shore. Some coastal buildings completely washed away. Evacuations will likely be ordered for areas susceptible to storm-surge flooding. Storm surge generally 13 to 18 feet. Central pressure 920–944 mb.

Category 4 hurricanes. Hurricane Charley when it hit Port Charlotte, Florida, in 2004. Hurricane Donna when it hit the Middle Florida Keys in 1960.

Category 5: Higher than 155 MPH—Catastrophic Damage

Complete roof failure on many residences and prefabricated buildings. Extensive damage to exposed glass on all large buildings. Some complete building failures. All shrubs, trees, and signs blown down. Complete destruction of mobile homes and poorly built buildings. Extensive window and door damage. Total destruction of all structures located near the shoreline. Storm surge generally greater than 18 feet. Central pressure less than 920 mb.

Category 5 hurricanes. Only three have hit the United States: the Labor Day Hurricane when it hit the Florida Keys in 1935; Hurricane Camille when it hit southern Mississippi in 1969; Hurricane Andrew when it hit Dade County, Florida, in 1992.

Other Considerations

Winds squared. The pressure from the wind increases with the *square* of the wind speed. Do the math and you'll find that a category 4 hurricane's winds exert more than twice the pressure on structures than a category 2.

Geometric damage increase. As the wind pressure increases, more things reach their failure point. For this reason, the damage geometrically increases with each Saffir-Simpson category. It's estimated that the damage goes up by four times with each higher category until you get to category 4. There is no effort to distinguish between category 4 and category 5 damage.

High-rise winds. Winds speeds are known to be significantly higher above the ground. This has serious implications for high-rises and mountainous areas. Winds can easily be a category higher at an elevation of 200 to 300 feet, that is, above the twentieth floor.

Domino effect. Buildings don't normally fail all at once. A weak door latch, poorly nailed roof panel, badly braced gable end, or some other weak spot normally starts the process. It's a domino effect. The trick is to keep the first dominos from falling. A building is only as strong as its weakest point.

History of the Saffir-Simpson Hurricane Scale

Herb Saffir, a Miami-area engineer, decided that hurricanes needed a scale to match the successful Richter scale used for earthquakes. He introduced the wind portion of the scale in 1971 in a report he prepared for the United Nations on low-cost housing in hurricane-prone areas. Robert Simpson, director of the National Hurricane Center at the time, added the storm-surge part of the scale after his experience with Hurricane Camille in 1969. The Saffir-Simpson scale was first used in hurricane advisories in 1975.

Z-Time, GMT, and UTC

Hurricane advisories are issued in a combination of the local time where the storm is at the moment, and Greenwich mean time, also called Z-time. Obviously, it's most convenient for people that might be affected by an approaching hurricane to get information referenced to their local time zone. But, hurricane forecasts are disseminated worldwide, so a standard time reference is also used, GMT or Z-time. This time standard was developed in the nineteenth century for exactly this reason, to be sure everybody everywhere knows exactly what time you're talking about.

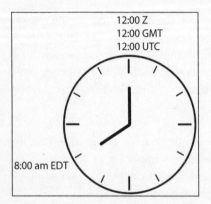

12:00 Z
12:00 GMT
12:00 UTC

8:00 am EDT

GMT history. As international travel and communications increased in the late nineteenth

century, the standardization of the labeling on maps became an imperative, and a standardized time system would become necessary for the same reason.

An 1884 international agreement set the meridian (the longitude line) running through Greenwich, England, a London suburb, as the prime meridian. This is where we start counting the number of degrees around the earth. So, for example, 80°W refers to the longitude line 80 degrees west of that prime meridian. Most of the navigational maps in the nineteenth century were British and already had the starting line through Greenwich, so it made sense to make it official.

But while there was agreement on map labels, there was no official international treaty on time zones, and there still isn't. It's left to each government. Time was made uniform by law in the United States in 1915, and by 1929 most major countries in the world had adopted the time zones that were implicit in the 1884 agreement.

There are twenty-five time zones around the world. The zone that straddles the International Dateline is divided in half because the calendar date is different on each side of the line. Letters are used to designate each zone, except *J* is reserved for local time.

A to Z time. The earth was divided into 15° slices (360° around the globe divided by 24 hours), creating the time zones. Each zone was assigned a letter beginning with the zone immediately to the *east* of the Greenwich-centered zone. What today we call the Central Europe time zone received the letter *A* (GMT + 1 = Zone A). The letter *J* was not used (it's used for the observer's local time), so GMT + 12, halfway around the world, was designated Zone M. Moving *west* from Greenwich, the zones were designated with the twelve letters *N* through *Y*.

The letter *Z* was reserved for "zero time" or Greenwich time. Thus, midnight in England is 0000Z. Eastern standard time is time zone R, GMT − 5.

NOTE: The above only applies to standard time. During the summer, EDT becomes GMT − 4, and British summer time becomes GMT + 1.

In the 1950s the International Civil Aviation Organization introduced a phonetic alphabet to be used to convey letters and numbers in aviation communications. Zulu is used for *Z* and Romeo is used for *R*, for example. That's why you sometimes hear Z-time referred to as Zulu.

Greenwich Mean Time is no longer, technically, the world standard, although in practical use there is no difference between today's Coordinated Universal Time and GMT. The current abbreviation is UTC (the abbreviation of the French spelling).

Understanding Hurricane Advisories

Official hurricane advisories are issued only by the National Hurricane Center in Miami. Other hurricane bulletins are issued by local National Weather Service offices, but they pertain only to the region served by the issuing office. Each office has responsibility for a group of counties called the County Warning Area. As you will see, a tremendous amount of information is available about any storm, and it's almost impossible to take it all in.

The schedule. Anytime there is a tropical depression, tropical storm, or hurricane active in the Atlantic, the National Hurricane Center issues a full "advisory package" every six hours: 5:00 A.M., 11:00 A.M., 5:00 P.M., and 11:00 P.M. eastern daylight time.

In practice the advisories come out somewhat before the official advisory time, although the Forecast Discussion often lags when a storm is threatening land.

Standard time schedule. Hurricane advisories from the NHC are issued at the same "absolute" time, not the same clock time, throughout the year. So, when the time changes to eastern standard time in the fall, the advisories come out an hour earlier: 4:00 A.M. EST, 10:00 A.M. EST, and so forth.

Synoptic times. So-called synoptic times are loosely based on the time that weather data is collected around the world, mainly 0000Z and 1200Z (see page 130, Z-time), which translates to 8:00 P.M. EDT

(7:00 P.M. EST) and 8:00 A.M. EDT (7:00 A.M. EST). Some data is also collected at 0600Z and 1800Z as well. These times—0000Z, 0600Z, 1200Z, and 1800Z—are called synoptic times.

The advisory schedule allows enough time for the data to be processed and the computer models to be run. Since the data come in an hour earlier during standard time, the advisories are issued an hour earlier as well.

National Hurricane Center Bulletins

Public Advisory

This bulletin is a *text version* of the latest information on the storm and is designed to be read by radio announcers and used by the general public. Detailed forecast information is *not* given in this bulletin. Instead, general statements are made about the expected future path and strength of the storm. Forecasts for storm-surge heights, rainfall, and the threat of tornadoes are also included. Also, recent wind or pressure readings of note are often listed. Wind speeds are given in mph.

Although the order is sometimes rearranged, the elements of the Public Advisory arc standard. Here we use the 5:00 A.M. EDT (4:00 A.M. CDT) advisory on Hurricane Wilma, on October 23, 2005, as an example:

Headline. An overview of the status of the storm.

. . . WILMA MOVING SLOWLY NORTHEASTWARD . . .
STILL NOT STRENGTHENING . . .

Watches and warnings. A section that can be quite lengthy on new, changed, remaining, and dropped watches and warnings.

Center location. The estimated location of the center of the storm at advisory time.

AT 4 AM CDT . . . 0900Z . . . THE CENTER OF
HURRICANE WILMA WAS LOCATED NEAR LATITUDE 22.1

```
NORTH . . . LONGITUDE 86.6 WEST OR ABOUT 40
MILES . . . 65 KM . . . NORTH OF CANCUN MEXICO
AND ABOUT 350 MILES . . . 565 KM . . . WEST-
SOUTHWEST OF KEY WEST FLORIDA.
```

Note that the time given on the advisory is local time in the area that is expected to be affected by the storm. Since Wilma was still near Mexico, the Public Advisory time is given in central daylight time, the local time at the storm's location.

Movement. The estimated direction and speed of movement of the center of the storm.

```
WILMA IS MOVING TOWARD THE NORTHEAST NEAR
3 MPH . . . 6 KM/HR . . . AND AN INCREASE IN
FORWARD SPEED IS EXPECTED DURING THE NEXT 24
HOURS.
```

Often there are qualifiers and averages in this section. A storm may have been moving to the west for the last two hours, but be expected to turn back toward the northwest. In that case, the wording may give the average over time or the expected average over the next few hours. Also, expected changes in direction will be mentioned in this section.

Highest winds. The estimated maximum wind speed of the storm and the Saffir-Simpson category.

```
MAXIMUM SUSTAINED WINDS ARE NEAR 100 MPH . . .
160 KM/HR . . . WITH HIGHER GUSTS. WILMA IS A
CATEGORY TWO HURRICANE ON THE SAFFIR-SIMPSON
SCALE. SOME INCREASE IN STRENGTH IS POSSIBLE
TODAY.
```

The maximum sustained wind speed is the highest wind (bearing in mind the definition of *sustained*, see page 85) that exists *anywhere* in the storm. Winds of that speed could exist in a large part of the eyewall, or only in one cell in a feeder band. See page 126.

Storm size. The estimated *maximum* radius of hurricane and tropical-storm-force winds from the center of the storm. For example, this from the 11:00 A.M. EDT/10:00 A.M. CDT advisory on Hurricane Katrina on August 28, 2005:

```
HURRICANE FORCE WINDS EXTEND OUTWARD UP TO 70
MILES . . . 110 KM . . . FROM THE CENTER . . .
AND TROPICAL STORM FORCE WINDS EXTEND OUTWARD
UP TO 200 MILES . . . 325 KM.
```

The distances given are the maximum distances of *any* hurricane or tropical storm force winds in any direction from the center. Most often hurricanes are asymmetric. For example, at the time of this advisory, tropical storm force winds were estimated to extend out 200 miles *only* to the east. At the same time they reached just 145 miles to the southwest of the center.

Lowest pressure. The estimated lowest barometric pressure at the center of the storm.

```
AN AIR FORCE HURRICANE HUNTER AIRCRAFT MEASURED
A MINIMUM CENTRAL PRESSURE OF 961 MB . . .
28.38 INCHES.
```

NOTE: The fact that this advisory says that the pressure was *measured* by Air Force Hurricane Hunters indicates that it is a credible number that is likely very close to exactly right. Often the advisory will say that the pressure was "estimated." This is a subjective estimate by the forecaster in light of the cloud pattern, previous pressure measurements, nearby observations, and any other information that might be available. These estimates are most often reasonably accurate, but occasionally the pressure is much higher or much lower than the clues would seem to indicate.

If a storm is close to making landfall there will be sections on:

Storm surge. Expected heights and areas most likely to be affected.

Rainfall. A range is given and often a maximum amount.

Swells. Areas that will likely be affected by large swells being generated by the storm.

Tornadoes. Where they are most likely to occur.

Repeat section. Near the end of the public advisory there is always a "repeat" section. It's a quick place to look to get an overview of the storm.

```
REPEATING THE 4 AM CDT POSITION . . . 22.1 N
. . . 86.6 W. MOVEMENT TOWARD . . . NORTHEAST
NEAR 3MPH. MAXIMUM SUSTAINED WINDS . . .
100MPH. MINIMUM CENTRAL PRESSURE . . . 961MB.
```

Next advisory. When the next Public Advisory will be issued, and, if applicable, when the next full advisory package will be issued.

Forecast Advisory

Embedded in this bulletin is the National Hurricane Center's *forecast* of where they think the storm is going and how strong it's going to be. Only the watches and warnings are written in plain English. The details of the current status and the forecast are given in a tabular format.

This bulletin is produced in a rigid format because it's often used as the source for computer-generated graphics. Web sites and television stations that present hurricane graphics get most of their information from this bulletin.

Distances are given in nautical miles (NM) and winds speeds in knots (KT). To convert to statute miles and miles per hour, respectively, multiply each by 1.151.

Times are all given in Z-time (see page 130). To convert to eastern daylight time, subtract four hours, or for central daylight time, subtract five hours.

Here are the components of the Forecast Advisory using the 5:00 A.M. EDT advisory on Hurricane Ivan from September 10, 2004, as an example:

Watches and warnings. Identical to the equivalent section in the Public Advisory.

Center location. The estimated location of the center of the storm at advisory time.

```
HURRICANE CENTER LOCATED NEAR 15.9N 74.02W AT
10/0900Z
```

Remember, 0900Z is 5:00 A.M. EDT. See page 130.

Center accuracy. A subjective estimate of the accuracy of the position.

```
POSITION ACCURATE WITHIN 10 NM
```

If there was a recent fix of the location of the center of the storm from a Hurricane Hunter aircraft, the number will be low, as this one is. If the cloud pattern is ragged and only satellite estimates are used, the number will be big.

Movement. The estimated direction and speed of movement of the center of the storm.

```
PRESENT MOVEMENT TOWARD THE WEST-NORTHWEST OR
300 DEGREES AT 11 KT
```

As in the Public Advisory, this is *not* the short-term motion of the center. It's better thought of as a representative motion over the recent past and the short-term future.

Pressure. The estimated lowest barometric pressure at the center of the storm.

ESTIMATED MINIMUM CENTRAL PRESSURE 930 MB

Eye diameter. The estimated diameter of the eye, if one is present and a recent measurement is available.

EYE DIAMETER 20 NM

Maximum winds. The estimated maximum sustained winds and gusts anywhere in the storm.

MAX SUSTAINED WINDS 125 KT WITH GUSTS TO 155 KT

Current wind radii. A block containing the estimated distance out from the center in the four quadrants that winds over *minimum tropical storm strength* (34 KT), *strong tropical storm strength* (50 KT), and *hurricane force* (64 KT) extend. Also the estimated distance from the center that the storm is creating seas of at least 12 feet is given. From a Hurricane Rita advisory:

```
64 KT. . . .  . . . 45NE 35SE 30SW 45NW.
50 KT. . . .  . . . 125NE 60SE 60SW 60NW.
34 KT. . . .  . . . 150NE 90SE 90SW 150NW.
12 FT SEAS. . . . . 360NE 140SE 140SW 200NW.
```

These numbers, given in nautical miles, are often rough estimates. Unless there happen to be buoys or ships in the storm, there is often no way to know for sure how far out the winds extend. Hurricane hunters, a special satellite called QuikSCAT (see page 173), and coastal Doppler radars can help, but they aren't available all the time. Sometimes forecasters have to make an estimate from satellite pictures based only on their experience and the deductions of the satellite analysts.

Notice that the hurricane is not symmetrical. Strong winds extend farther out to the northeast than any other direction. This is almost always the case.

Repeat section. A repeat of the advisory-time estimated location of the center plus an estimated location of the center at the most recent synoptic time (see page 129).

```
REPEAT . . . CENTER LOCATED NEAR 15.9N 74.2W AT
10/0900Z AT 10/0600Z CENTER WAS LOCATED NEAR
15.7N 73.8W
```

NOTE: The synoptic time (0600Z or 2:00 A.M. EDT in this case) location of the center is not especially significant while the storm is happening. No one is very concerned with where a storm was three hours ago. But the forecasts are all based on synoptic times, not advisory time. The twelve-hour forecast in this advisory will be for 1800Z or 2:00 P.M. EDT. So, to determine the twelve-hour forecast movement of the storm, and for verification of the forecast after the fact, it is important.

Forecast. This is the only place in the whole package of bulletins and graphics that the full body of forecast information on the storm is given. There are forecast locations for the storm's center, maximum sustained winds, gusts, and maximum wind radii in the same format as above.

```
FORECAST VALID 10/1800Z 16.9N 75.6W
MAX WIND 125KT . . . GUSTS 155KT.
64KT . . . 45NE 35SE 35SW 45NW.
50KT . . . 100NE 60SE 60SW 60NW.
34KT . . . 150NE 90SE 90SW 150NW.

FORECAST VALID 11/0600Z 18.0N 77.4W
MAX WIND 130KT . . . GUSTS 160KT.
64KT . . . 45NE 35SE 35SW 45NW.
50KT . . . 80NE 60SE 60SW 60NW.
34KT . . . 150NE 90SE 90SW 150NW.

FORECAST VALID 11/1800Z 19.1N 78.9W
MAX WIND 135KT . . . GUSTS 165KT.
64KT . . . 45NE 35SE 35SW 45NW.
```

```
50 KT . . . 80NE  60SE  60SW  60NW.
34 KT . . . 150NE  90SE  90SW  150NW.


FORECAST  VALID  12/0600Z  20.4N  80.2W
MAX  WIND  135 KT . . . GUSTS  165 KT.
50 KT . . . 80NE  60SE  60SW  60NW.
34 KT . . . 150NE  90SE  90SW  150NW.


FORECAST  VALID  13/0600Z  23.3N  81.9W
MAX  WIND  115 KT . . . GUSTS  140 KT.
50 KT . . . 80NE  60SE  60SW  60NW.
34 KT . . . 150NE  90SE  90SW  150NW.
```

EXTENDED OUTLOOK. NOTE . . . ERRORS FOR TRACK HAVE AVERAGED NEAR 250 NM ON DAY 4 AND 325 NM ON DAY 5 . . . AND FOR INTENSITY NEAR 20 KT EACH DAY

```
OUTLOOK  VALID  14/0600Z  27.0N  82.5W
MAX  WIND  115 KT . . . GUSTS  140 KT.


OUTLOOK  VALID  15/0600Z  31.0N  82.5W . . . INLAND
MAX  WIND  50 KT . . . GUSTS  60 KT.
```

Notice several things. First, the time for which each forecast is valid.

Date/Time:	10/1800Z	11/0600Z	11/1800Z	12/0600Z	13/0600Z	14/0600Z	15/0600Z
Forecast Period:	12 hours	24 hours	36 hours	48 hours	72 hours	96 hours	120 hours

Forecasts are made for the state of the storm 12, 24, 36, 48, 72, 96, and 120 hours after the most recent synoptic time—in this case, 0600Z or 2:00 A.M. EDT—not the issuance time of the advisory.

Forecasts are made at 24 hours intervals after 48 hours because projections get less precise as you go farther into the future.

The 96- and 120-hour forecasts, in this case, for 14/0600Z and 15/0600Z, respectively, are separated from the other time periods to indicate that the average track and intensity errors are quite large.

The 120-hour forecast for 15/0600Z projects that the storm will have crossed the coast. When this is the case, INLAND is included on the location line. Also, notice the forecast intensity at that time was down to 50 knots.

Next advisory. The time the next Forecast Advisory will be issued.

NEXT ADVISORY AT 10/1500Z

Forecast Advisories are scheduled to be issued four times a day. Unlike the Public Advisory, the frequency does not increase as storms get closer to the coast. Therefore, even though *current* storm information is updated more frequently (see page 132), the *forecast* is only updated every six hours. Occasionally a special advisory is issued, but it's rare.

Forecast Discussion and Other Bulletins

Following is a list of other bulletins issued by the NHC. Examples and detailed explanations of each are available at www.hurricanealmanac .com.

Forecast Discussion. The Forecast Discussion provides a *technical explanation* of the rationale behind the NHC forecast. Included in it is the forecaster's analysis of the validity or lack of validity of the data available to him. There is usually an explanation of the various computer models, the trends they indicate, and whether they are individually or as a group credible.

Originally designed for internal government communications, these Discussions have become more user-friendly over the years. However, they occasionally still venture off into fairly deep technical waters. There are only two sections to this part of the advisory package: the text section and the forecast section.

The length of the Discussion varies tremendously, depending on the current or potential threat to land the storm is posing and the level of confidence the forecaster has in his ability to make an accurate forecast with the amount of data that's available.

Often the forecast uncertainty is related to the degree of consensus in the computer models. When there is reasonable agreement in the major models, the forecasters normally feel more confident about the forecast.

The forecast locations are the same as those given in the Forecast Advisory. The predicted wind speeds are given in knots.

Probabilistic Surface Wind Speed Text Product. This product, with the ridiculously unwieldy name, was new for the 2006 hurricane season. It replaced the old Strike Probabilities product, which was an extremely important, but little read, part of the advisory package.

The new product gives the probability of winds of a certain strength affecting a given location, whereas the old product only gave the odds of the storm's center coming within 75 miles of the location. It said nothing about the size or strength of the storm and what the effect might be. A small hurricane's center could be 75 miles offshore with little effect on the coast, while a large hurricane could be having a major impact. The new bulletin attempts to address that issue.

While the old product was based on the NHC's forecast errors over the past ten years, the new numbers come from 1,000 possible storm tracks simulated by a computer; the number of times winds of tropical storm or hurricane force occur at given locations determine the wind probabilities.

National Hurricane Center Graphics

The Forecast Cones

Officially called the Watch/Warn 3-Day and 5-Day graphic, this is the National Hurricane Center's version of the famous *forecast cones*.

Making the cone. The cone graphic is made by first drawing a circle around each of the seven NHC forecast positions for the center of the storm from the Forecast Advisory.

The radius of each circle is equal to the average error in NHC forecasts for that time period over the previous ten years. For example, the average NHC 24-hour forecast error in use in 2005 was 87 miles, so a circle of that radius was drawn around 24-hour forecast point. Similarly, an appropriate-radius circle is drawn around each of the forecast points. The edges of those circles are then connected together to form the cone.

Cone-graphic elements. This Watch/Warn 5-Day graphic was issued for Tropical Storm Rita at 11:00 P.M. EDT on Monday, September 19, 2005. The important components are:

- The solid white 3-Day Cone
- The white hashed 4- and 5-day extension of the cone
- The watch and warning colors in Florida, the Bahamas, and Cuba

Additionally, the NHC includes a center black line connecting the forecast positions. This line is *extremely misleading* and has caused significant misunderstandings in the past. The odds of the center of the storm deviating significantly from this line are essentially equal to odds of the storm following the line.

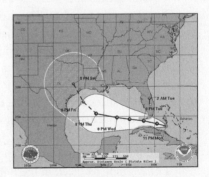

The famous forecast cone graphic from Tropical Storm Rita, September 19, 2005. It's sometimes affectionately called the Cone of Death.

LOOK AT THE CONE, NOT THE LINE.

Cone Holds the Center About 70 Percent of the Time

The center of the storm stays inside the cone about 70 percent of the time. You might think it should be 50 percent since the cone width is set by the "average" track errors, but there are also errors "along" the track. Said another way, some forecasts have the time of arrival wrong, but the storm still goes in the forecast direction, which means the center stays within the cone. So the cone is a good guide to where the storm is going.

3-Day Cone. This is the most important part of the graphic. Very often, when the 3-Day Cone touches land, early preparations need to begin for the possibility of a storm strike. When the cone first touches the coast, in most cases, there is plenty of time to prepare. Full-scale preparations are normally not called for, but your plan for the next two to three days needs to be finalized.

5-Day Cone. This part of the graphic is, in my opinion, often more inflammatory than useful for most of the population. That's mostly the media's fault because the 4- and 5-day parts of the cone are often colored and treated exactly like the 3-Day Cone. That's misleading. The fact that your area has been touched by the shaded part of the graphic (the 4- and 5-day area) means very little. The graphic is over 700 miles across at its widest point, so much of the area in the wide part of the cone is likely to feel very little effect from the storm. If your part of the coastline is touched by the shaded area, it is time to go over your hurricane plan and be ready to take more action in a couple of days if the storm keeps coming.

Watch and warning areas. In theory, these are the areas under a watch or warning as listed in the Public Advisory. However, beware of using this graphic as the last word. I've seen numerous instances were the colored areas were slightly different than the actual part of the coastline put under a watch or warning. It's a reasonable guide, but use the text advisory as the final word.

New   **Old**

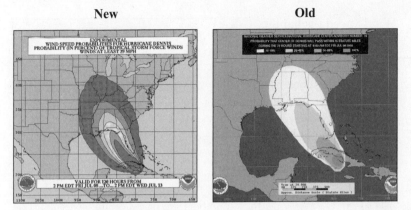

Two graphics from Hurricane Dennis, July 8, 2005. The new graphic shows the odds of 39 mph+ winds over the next five days. The old graphic shows the odds of the center of the storm coming within 75 miles of a given location.

Wind Speed Forecast and Probabilities Graphic

This new graphical product, on the left, replaces the strike probabilities graphic on the right. The old product only dealt with where the center of the storm might be over the next three days. The new graphic gives a sense of where the storm is forecast to go as well, but the goal of the product is to communicate what areas might experience winds of at least *tropical storm* force.

The background data for the new graphic is the Probabilistic Surface Wind Speed Text Product (see page 164). While the text product gives probabilities for three strengths of wind, including hurricane force, this graphic *only* shows the odds of tropical storm force winds or higher occurring during the next five days.

In the graphic on the left, from Hurricane Dennis in July 2005, the dark color near the center indicates a 90 percent chance of winds of 39 mph or more. The dark color at the edge of the graphic represents the area where the odds are between 5 percent and 20 percent.

Notice that the new graphic is wider than the old one because the old graphic was concerned only with the center of the storm. The new graphic includes the effect of tropical storm force winds that extend well away from the center.

Wind Speed Probability Table

This graphic is to wind speeds as the cone is to the forecast track. The graphic's goal is to provide a sense of the uncertainty in the National Hurricane Center's forecast intensities. As with all of the probability products that are part of NHC advisories, these numbers are based on the average errors in past forecasts. There is nothing subjective about the numbers in this table.

WIND SPEED FORECAST FOR IVAN
EXPRESSED AS PROBABILITY
FROM NHC ADVISORY 52
4:00 AM CDT SEP 15 2004

TIME	WIND SPEED INTERVAL IN MPH							
HOURS	DISSIPATED	TROPICAL DEPRESSION < 39	TROPICAL STORM 39 - 73	HURRICANE >= 74	HURRICANE			
					CAT. 1 74 - 95	CAT. 2 96 - 110	CAT. 3 111 - 130	CAT. 4-5 >= 131
12	<2%	<2%	<2%	>98%	<2%	3%	40%	55%
24	<2%	<2%	<2%	>98%	<2%	10%	40%	50%
36	<2%	2%	35%	65%	45%	15%	5%	<2%
48	20%	35%	35%	10%	10%	<2%	<2%	<2%
72	30%	30%	25%	15%	10%	<2%	<2%	<2%

It's pretty straightforward once you get the hang of it. As an example, here's a Wind Speed Probability Table from the 5:00 A.M. EDT/ 4:00 A.M. CDT advisory on Hurricane Ivan on September 15, 2004:

Consider the top line, the 12-hour forecast. The Forecast Advisory from the same advisory package as the table predicts that Ivan will have 140 mph winds (category 4) in 12 hours. The table says:

- There's a 98 percent chance that Ivan will still be a hurricane.
- There's a 55 percent chance that Ivan will still be a category 4 or 5.
- There's a 40 percent chance that Ivan will weaken to a category 3.
- No other category of strength has a significant likelihood of occurring.

Consider the 36-hour line. The Forecast Advisory predicts that Ivan will have 80 mph winds in 36 hours. The table says:

- There's a 65 percent chance that Ivan will still be a hurricane.
- There's a 35 percent chance that Ivan will have weakened to a tropical storm.
- There's a 45 percent chance that Ivan will be a category 1 hurricane.

- There's a 15 percent chance that Ivan will be a category 2 hurricane.
- There's a 5 percent chance that Ivan will be a category 3 hurricane.
- No other category of strength has a significant likelihood of occurring.

Lastly, consider the 48-hour line. The Forecast Advisory predicts that Ivan will be inland at this time and have weakened to a tropical depression. The table says:

- There's a 10 percent chance that Ivan will still be a category 1 hurricane.
- There's a 35 percent chance that Ivan will still be a tropical storm.
- There's a 35 percent chance that Ivan will be depression as forecast.
- There's a 20 percent chance the system will have dissipated over land.
- No other category of strength has a significant likelihood of occurring.

The strength of the storm and the track are, obviously, tied together. If the storm moves slower than forecast, it would still be over the Gulf in forty-eight hours, for example, and likely be much stronger. So, track issues are key to intensity uncertainty. In addition, the science of predicting storm intensity is not as well developed as it is for track forecasts. So, taking everything into account, intensity forecasts must be considered estimates at best.

NOTE: For extremely strong storms, you may frequently see "NA" or "TF" displayed. This is because the program that fills the table looks for analogous hurricane forecasts in the historical record to make its probability estimates. This graphic is based on the years 1988–1997 (as opposed to the cone, which is based on the last ten years). During that period there were a small number storms with extreme wind speeds. Therefore, for storms forecast to reach near or above 150 to 160 mph, there will not be enough analogous cases, so the program can't run. In that case the box is filled with NA (not available) or TF (too few cases).

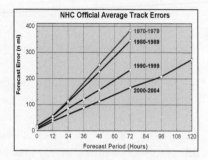

NOTE: The scale on the left side of the chart is in "n mi" or nautical miles. To convert to statute miles, multiply by 1.151. So 100 n mi = 115 miles, for example.

NHC Forecast Errors

Forecasts issued by the National Hurricane Center are getting better, on average, every year. The graphic above shows the improvement decade by decade since the 1970s. Still, there is a significant amount of average error in the forecasts in every time period, especially when you think about the dramatic difference a storm tracking a hundred miles offshore versus making landfall makes to residents of a coastal city.

People want to know whether they have to go into full-scale preparation mode or not. This chart's message is:

MORE OFTEN THAN NOT, YOU WILL PREPARE AND END UP NOT GETTING HIT.

Changes in 2006. Starting in 2006 the National Hurricane Center based its "average" error on the past five years, instead of the past ten. Below is the chart of average NHC errors by time period from 2001 to 2005. Obviously, the running five-year average changes slightly after each hurricane season as new forecast errors are added in. Check www.hurricanealmanac.com for the updated numbers.

	Initial	12HR	24HR	36HR	48HR	72HR	96HR	120HR
Error in miles	9	43	75	100	136	197	266	349

Each Forecast Advisory issued has a predicted latitude and longitude for the center of the storm at each of these time periods. The chart tells you how far off, on average, the positions for each time period have been over the five-year period. All tropical depressions, tropical storms, and hurricanes in the Atlantic, Caribbean, and Gulf of Mexico are included in the average.

So, to be absolutely clear, because this is important, let's imagine a 5:00 P.M. EDT Forecast Advisory. Each position, whether initial or forecast, has an *intrinsic* error associated with it. Reading the chart from left to right:

- The initial position of the storm is where the National Hurricane Center estimates it *was* at the previous synoptic time, which in this case was 2:00 P.M. EDT (or 1800Z, see page 130 for an explanation of Z-time). On average between 2001 and 2005 the NHC was off by an average of 9 miles as determined by post-storm analysis.
- The 12-hour forecast position—where the NHC thought the center would be at 2:00 A.M. the next day—given in the advisory was off by an average of 43 miles.
- The 24-hour position—where the NHC thought the center would be at 2:00 P.M. the next day—was off by an average of 75 miles. And so forth.

So when evacuations get under way in a given location twenty-four hours or more before the center comes ashore, the average error is about 75 miles. That's why preparations and evacuations are often undertaken when they end up not having been necessary. Unfortunately, it will always be that way.

Understanding Average Errors

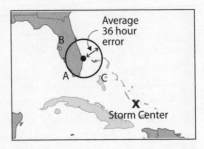

Consider this scenario: You are living along the coast where the dot is on the map. The center of the hurricane is offshore and is forecast to head in your direction. In fact, the National Hurricane Center's Forecast Advisory has the expected location of the center of the storm in thirty-six hours over your town. Yikes!

The circle represents the average error in the NHC forecasts. If this storm is forecast like an average storm, 50 percent of the time the center of the hurricane will be inside the circle in thirty-six hours and 50 percent of the time it will not be. Correct? Yes, that's what average means. (NOTE: The center stays inside the cone about 70 percent of the time because a storm that goes significantly faster or slower than forecast, but stays near the center line, would stay inside the cone, but could end up outside the thirty-six-hour error circle, for example.)

If you don't think carefully, these numbers can lead you to misunderstand your risk. Think about the locations A, B, and C outside the circle. Consider what your risk would be if the storm center ended up at one of those points in thirty-six hours:

A: You would likely have *little effect* from the storm.

B: It's likely *you were hit* by the storm.

C: There's a reasonable chance that *you will still be hit* by the storm.

The point is, not all errors are across the track, left to right. In fact, errors are slightly more likely to be *along* the track, which are errors in time rather than in where the storm is going.

The bottom line is, if you are inside the circle around a 12-, 24-, or 36-hour forecast position—that is, if you're inside the cone—the odds are high enough that you must prepare.

Special Satellites

QuikSCAT

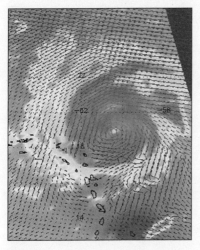

QuikSCAT image from Hurricane Frances on August 31, 2004. Courtesy NASA.

An instrument called the Sea-Winds Scatterometer riding on NASA's QuikSCAT satellite measures surface winds over the ocean. Two radar pulses are sent from the satellite to the ocean surface. When the waves are big and there's a lot of spray, a strong signal is reflected back to the Sea-Winds system. A computer then analyzes the return signals from the two beams and calculates an estimate of how strong the winds are at that location and what direction they are blowing. The wind speed data is only valid in rain-free areas, and it has been found that the system tends to underestimate the winds in strong hurricanes, however.

The SeaWinds/QuikSCAT system has given forecasters good information on the size of storms well away from land for the first time. Previously, the extent of tropical storm force winds, for example, was often a guess.

The QuikSCAT satellite is a polar orbiter, so it only covers a given spot on the earth at most twice a day.

AMSU—Advanced Microwave Sounding Unit

This instrument measures temperatures at various levels in the atmosphere. It's used as a research tool, to measure global warming, for example. But it can be useful in helping to determine whether an extratropical or subtropical storm has gained tropical characteristics. Forecasters will look at the temperature of the core of the system. If it has warmed, they take that as a sign that the system is becoming a tropical cyclone.

The instrument rides on the NOAA-15 polar-orbiting satellite, so data is normally available twice a day.

5

How I'd Do It Better

Some Thoughts

In 1983, I remember watching television coverage of huge sheets of glass from Houston's downtown high-rises slicing through the air, filling the streets with smashed windows and piles of dangerous debris. It was caused by Hurricane Alicia's winds blowing roof gravel—standard roofing material at the time—off adjacent buildings into neighboring skyscrapers, shattering the glass walls and windows. At the time I thought, "What idiot didn't think that was going to happen in a hurricane?"

Over the last twenty-three years—and especially during 2004 and 2005—we've had hurricane after hurricane come ashore with devastating, but predictable, results. We've seen every aspect of the hurricane system fail, from communications to protection to recovery. But, still, precious little has been done to significantly change the systems that have been in place for decades.

What follows are my thoughts on how to modify or completely change several components of the "hurricane system." I'm concentrating here on property protection (and how insurance plays into it), communications, and evacuations because, in my opinion, they are the most important issues and can make the most difference. How we recover

from storms is also critical, but if we do a better job on the front end, recovery becomes much easier.

These ideas are *not* comprehensive. They are starting points for a
conversation, areas where I think research could make a lot of difference. In some cases there are similar or related programs being talked
about or in effect. But none of the proposals I've seen go far enough toward real solutions. We need a concerted effort to change the deficient
aspects of the systems we use to deal with hurricanes, and we need
it now.

One more thing I want to make clear. I'm going to be critical in this
section of some of the policies and procedures of the National Hurricane Center. However, I am in *no way* being critical of the people who
work there. They are, in my opinion and experience, as fine examples of
public servants as can be found in the government. They work tirelessly,
along with many other people in the National Weather Service, to provide the best information possible. The system they work with, however, is antiquated and not in step with today's communications
realities. In my opinion, it's a system in need of updating.

From using words that confuse people, to protecting a city below
sea level with a levee system known to be deficient, to refusing to adopt
logical and reasonable policies so people have affordable insurance, our
government has let us down. *Predictable* bad things keep happening,
and many of us keep saying, "What idiot didn't think that was going to
happen in a hurricane?"

The Insurance Crisis

The Only Solution to the Insurance Crisis: A National Catastrophe Fund

The hurricane insurance problem is a crisis that's about to turn into
a catastrophe in its own right. *No* insurance system can *ever* work for
hurricanes because:

- Hurricanes don't happen with enough regularity so a reliable average yearly loss number can be calculated.

- No matter how big a pile of money an insurance company has on hand to handle a major hurricane, it may not be enough if there's a monster storm.
- There could always be another monster storm next week.

This is in contrast to fires and car accidents, for example, which happen with enough uniformity that calculations can be made to determine what an appropriate premium needs to be for the company to pay its claims and make a reasonable profit. These high-frequency events are "insurable" because they occur with predictable regularity.

Major hurricanes, devastating earthquakes, and serious terrorist attacks, that is, catastrophes, only rarely happen and therefore don't fit the definition of an insurable event. Every time there is a hurricane or a series of hurricanes, rates go higher. They have to, because no amount of money is guaranteed to be enough.

Hurricane insurance is really a misnomer. When you pay money to the insurance company to cover you against a hurricane, you are really just throwing money into a pot and hoping it's going to be enough. Even though it is *guaranteed* that this year, next year, or another year soon it will not be enough. The system will collapse and the federal government will have to come bail us out to the tune of billions and billions of dollars.

How I'd Do It Better

There is only one solution to the problem: a National Catastrophe Fund. By taking the relatively rare extreme event out of the insurance equation, insurance companies can make a reasonable estimate of what their maximum yearly losses will be and what they have to charge.

There are a variety of ways to implement this program, but the key component of any properly constituted catastrophe fund system must be *incentive*. Communities and states should be allowed to participate in the fund—and therefore have their insurance rates dramatically lowered—only if some or all of the following conditions are met:

- The building code meets a high standard (appropriate to the local threat) and is properly enforced.

- A program is undertaken to retrofit existing buildings.
- A program is undertaken to strengthen the homes of people who, for economic reasons, are unable to participate in other programs (many of whom likely carry no hurricane insurance under the current system).
- New mortgages are not granted unless the building being sold is brought up to a realistic standard.
- A practical system is created to "rate" individual buildings for loss potential.
- The state creates an escalating insurance scale based on a building's "rating."
- The state has its own catastrophe fund.

The fund would cover *all hurricane losses* above a defined amount *each year* to protect insurers from multiple-storm seasons. Since total payouts for any one season are capped, insurance would become more widely available and competition would return to the marketplace.

Many people have a philosophical problem with the government's being in the insurance business. The fact is, however, currently insurance companies are *not in the insurance business* when it comes to hurricanes.

A properly conceived National Catastrophe Fund facilitates the creation of a robust and realistic "National Catastrophe Protection System." And it's the only practical way to get that done.

The other complaint against creating a catastrophe fund on a federal level goes something like "People in North Dakota shouldn't have to pay for losses caused by people living in the way of hurricanes." I have four responses to that idea:

- People in North Dakota and every other state are going to pay anyway. Look at the federal bill for Katrina, Rita, Wilma, Andrew, Hugo, and every other bad hurricane—many billions of dollars.
- Each state should be required to have a state catastrophe fund. The size of that fund would vary depending on the likelihood of a devastating event happening in that state. Homeowners in Florida and California, for example, would have a significantly bigger

catastrophe fund surcharge on their policies than homeowners in Ohio. In case of a catastrophe, the state fund would be used up first.

- By creating the incentive system described above, the future losses for everybody are diminished, so there is the potential to save federal money down the road.
- The National Catastrophe Fund doesn't cost much of anything if there is no mega event. But, all the while it's providing the incentives necessary to improve the national readiness for a major hurricane, earthquake, or other horrific disaster.

As I said earlier, hurricanes are not, by definition, insurable. Add to that the economic drag high insurance rates can create, and there's even more reason to take action. The money paid out *after* a disaster only solves problems caused by that one event, a terribly inefficient system. Money spent on mitigation to minimize losses protects against all storms to come. Lives will be saved, cleanup costs for garden-variety disasters will be dramatically diminished, and we will make a better country.

There is, of course, a federal "insurance" system in place, the National Flood Insurance Program. This flawed program should be reworked and incorporated into the new plan. Currently, the federal government is paying people to rebuild homes in locations that are guaranteed to be devastated by flooding again. It's a losing proposition. People that own property on vulnerable barrier islands should, in many cases, be allowed to rebuild, but not without meeting high standards. In some cases, however, there is no way to build a house high enough to make it safe from storm surge or other flooding. In that case, no federal (i.e., public) money should be paid to landowners wanting to reconstruct their homes. People who want to rebuild in these areas should do so with their own money at risk.

Everybody in the system, from insurance companies to informed legislators, knows that there is no way to stop the escalating insurance crisis in the coastal areas by simply moving piles of money around. Concrete steps must be taken to lower the risk. The sooner Congress moves on the program outlined above, or something similar, the better for the entire country.

Make Retrofitting Existing Buildings Affordable with Creative Financing

The biggest impediment to a homeowner's adding shutters and other hurricane protection to a house is, obviously, the cost. Even if the National Catastrophe Fund works as it should, significantly lowering insurance costs for protected properties, many people will still not be able to afford the expenditures necessary to make their properties hurricane resistant.

How I'd Do It Better

In the discussion of how to create an overall system of hurricane preparedness, one of the interested parties is often left out: the banks and other lending institutions. Mortgage holders should be interested in protecting properties and neighborhoods from catastrophic damage from hurricanes because their business is based on the value of the property. If property values drop, banks stand to lose. If the banks help protect their customers' properties, they protect themselves as well.

For many homeowners the problem can be solved by the creation of a "disaster protection home mortgage add-on" incremental loan system. The idea is to "add on" to an existing mortgage the money needed to buy shutters and other hurricane protection products and spread the cost over the remaining term of the mortgage.

For example, take a homeowner with $150,000 and twenty-seven years remaining on the mortgage on a $200,000 home. Say hurricane protection costs $10,000. The add-on program would allow that money to be added to the mortgage balance and spread over the full twenty-seven years, increasing the monthly payment about $60 to $70, depending on the interest rate. That would be affordable for most folks.

When combined with the national catastrophe fund that would dramatically lower insurance rates for protected properties, an add-on might actually save homeowners money in the end. But, even if they don't save much, it's a win-win-win situation.

- Homeowners will have a protected home and save on insurance in the future.

- Banks will reduce their exposure to losses from holding mortgages on destroyed properties in destroyed neighborhoods.
- The community will benefit from having better protected homes in which people can stay during a storm and be safe.

The National Catastrophe Fund could also guarantee the incremental increase in the loans so that credit issues don't stand in the way of hurricane protection.

Shelters Required

Require That Mobile Home Parks Be Safe for Residents

Many tens of thousands of people in the hurricane zone live in mobile homes. In many cases, of course, they do it for financial reasons. But that's not always the case. Some people like the camaraderie of some of the parks, and some prefer the similarity to a single-family home.

Modern mobile and manufactured homes are comfortable and much safer than the older styles, but they are still no place to ride out a hurricane. The alternatives are difficult, however, for many residents who have no good place to go. Many people, especially elderly folks who have to take medication or have other problems, are very reluctant to go to a shelter.

It's a dangerous situation that has been crying out for a solution for as long as there have been mobile homes.

How I'd Do It Better

In my opinion, any developer who congregates people, especially elderly people, in a place or facility that is dangerous during a hurricane has a moral—and maybe it should be legal—responsibility to provide a reasonable way for those people to be safe. In the case of a mobile home park outside of the evacuation zone, that means an in-park shelter.

This isn't even a complicated issue. Many parks already have clubhouses or community halls. The requirement should be that these facilities must be hardened, meeting a building standard like other hurricane shelters. Parks that don't have clubhouses should be required to construct a no-frills hardened structure that would be a safe haven for residents.

Require Coastal Condo Developers to Take Some Responsibility for Sheltering

In many parts of the hurricane zone, apartments and condos are being built near the water at a mind-boggling pace. Coastal living with nice breezes and sunset walks on the beach is great, until a hurricane comes along. More people living in evacuation zones aggravates an already overburdened evacuation and sheltering system. The problem is exacerbated by the fact that many of the new residents are from out of the area, often from "up north," and don't have established friends in their new hometown. This means they aren't likely to have an obvious place to go that is safe in a storm.

The current building frenzy will result in placing many more people in harm's way, when as evidenced by the hurricanes of 2005, the government is unable to deal with the situation we have now.

How I'd Do It Better

The same argument I used for mobile homes applies here. Part of the cost of congregating people in harm's way should include providing safe shelter. In the case of coastal apartments and condos, safe shelter can only be found outside the evacuation zone.

This doesn't have to be as hard as it sounds. Coastal developers often build projects away from the coast as well. Additional safe space could be provided in those developments for the people coming from the coast. Alternatively, coastal developers could meet their responsibilities by retrofitting and hardening inland buildings can be additional shelter space for the community.

The concept of "impact fees" is not a new one. Developers in many areas are responsible for schools, roads, and other public infrastructure that their new residents will require. This basic concept needs to be extended to hurricane sheltering.

Communications

Rethink How Hurricane Advisories Are Structured

The Problem: The National Hurricane Center knows a lot more about an approaching hurricane—what's it's likely to do, and who is at risk—than most of the general public ever understands. People are *always* surprised by what a hurricane does, while the experts at the National Hurricane Center rarely are.

The solution is complex because the problem is complex. Let's start with the basics, the way hurricane information is disseminated. The NHC breaks a fundamental rule of communications with the structure and design of the advisory system.

Bryan's Communications Rule 1: *The communications system must not modify the message.*

How does this rule apply to hurricane advisories? To convey the forecast location of the center of a hurricane, the NHC specifies a latitude and longitude. For example, just to pick one of the thousands of forecasts made in 2005: The Hurricane Rita Forecast Advisory issued Thursday, September 21, 2005, said that the center of the storm was *forecast* to be at latitude 27.0°N, longitude 94.5°W at 1:00 CDT on September 23, the so-called "48-hour forecast position."

Well, the forecasters didn't really think the storm was going to be *exactly* there, just somewhere *around* there—plus or minus 100 miles or so. So why give a *specific* quantity—the latitude and longitude—to communicate an *uncertain* quantity—where the center of the storm is going to be in the future?

The communications system of enumerating a specific location *modifies the message*.

The communicated message is: *The storm's center is likely to be at this defined spot in 48 hours*.

The intended message is: *The future location of the storm's center is uncertain, but in 48 hours it's most likely to be in this defined zone about 200 miles across*.

How I'd Do It Better

Instead of communicating the forecast location of a storm by a specific point, issue the boundaries of a *threat zone* for each time period. The 12-hour, 24-hour, 36-hour, 48-hour, 72-hour, 96-hour, and 120-hour threat zones plotted all together would create the same well-known forecast cone that we see today. But there would be no tendency to put the evil line down the center because there would be no "point" forecasts.

The tornado people in the National Weather Service figured this out a long time ago. They don't communicate where a tornado might hit later in the day by naming a specific neighborhood that seems most likely to get run over. They put out a risk area. Based on the best science available, what zone has the highest threat? The hurricane situation is similar enough to use the same system.

As research into creating an objective "confidence level" of a given hurricane forecast matures, the size of the threat zones could be dynamic, that is, variable. A low confidence level would mean a bigger threat zone, and vice versa. There's a lot of work to be done in this area, possibly involving development of a technique for objectively measuring the degree of consensus of the hurricane models.

The Curse of the W-Words and How the Warning Language Could Be Better

The fact that the National Weather Service uses the English words *watch* and *warning* to communicate the two levels of readiness for a severe weather event is only the beginning of the problems with the American severe-weather alert system. In my decades as a broadcaster I have heard news anchors, weathercasters (including me!), and directors of the National Hurricane Center say *watch* when they mean *warning* and vice versa. *The two words are confusingly similar*.

Ask the average person and they rarely know the difference. It doesn't make any sense to use confusing words to alert people to dangerous conditions. Am I missing something here?

Since this book is about hurricanes, let's just concentrate on the words used to warn for tropical systems. As you know, if you read the earlier part of the book, a *hurricane watch* is issued if hurricane conditions are *possible* within the next thirty-six hours or so, while a *hurricane warning* is issued when hurricane conditions are *probable* or *expected* within about the next twenty-four hours.

There is another *intrinsic* alert issued by the National Hurricane Center, however. When the forecast cone touches the coast or points right at it, people feel threatened. Yet there is no formal word for the obvious fact that there is a heightened level of risk in the targeted coastal area. When the forecast cone first touches the coastline, it's like an eight-hundred-pound gorilla just entered the room. But the NHC doesn't acknowledge it in a formal way because the level of risk doesn't meet the criteria necessary for a watch to be issued.

How I'd Do It Better

I think the tornado people have the right idea with three levels of notice that there is a threat of a severe event. I'd do that for hurricanes as well. So my system would be:

- About 72 hours before landfall: Define a *risk area* for the part of the coast that needs to pay close attention. The appropriate threshold would need to be determined, perhaps based on the new wind-speed probability product, but a three-day forecast cone on or near the coast would have to be acknowledged. This "risk area" would be a coastal zone, like a warning area, and would replace the 5-Day Cone (see next page).
- If the storm is expected to be especially strong or the forecast confidence is especially high (more research is needed in this area), an area of *high risk* would be identified.
- At 36 to 48 hours before an expected landfall, a *hurricane alert* would be issued, replacing the confusing *hurricane watch*.
- At about 24 hours before an expected landfall, a *hurricane warning* would be issued, as it is today.

I think this system would solve many of the confusion-caused breakdowns that occur today in communicating the status of the threat from a storm. Another benefit would be that areas like the Florida Keys that need to evacuate early (before a hurricane watch is issued) would be in a risk area, helping emergency managers motivate people to get moving.

Get Rid of the 5-Day Cone

The evidence is in. The 5-Day Cone does more harm than good. The harm comes in at least two forms: the unnecessary anxiety that is caused due to the expansive area the cone covers, and the damage done to the National Hurricane Center's forecast credibility because large numbers of people *feel* alerted when they are "in the cone," but then the storm never comes anywhere near them.

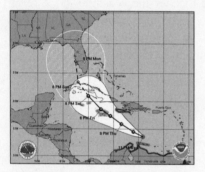

The 5-Day Cone issued by the NHC for Hurricane Ivan on September 8, 2004. The graphic implicitly alerted the entire peninsula and most of the panhandle of Florida. With the storm still well south in the Caribbean, it was too soon to raise any concern.

The 5-Day Cone is nearly 700 miles across at its widest point, bigger than the entire state of Florida. That's too big an area and places too many people unnecessarily under the pressure of a potential storm for too long.

The problem is aggravated since, for the most part, action is not required four or five days in advance. Hurricane plans need to be designed to be implemented at most 48 hours before a storm. For 99 percent of businesses and members of the public, there is no useful information in a cone that is issued five days before a *potential* strike, bearing in mind that the odds of the strike by the worst part of the storm are very, very low.

The 3-Day Cone, on the other hand, has significant value. If a coastal section is "under the gun"—that is, the cone is aimed at it—people there should begin preparations, at least mentally. At the 72-hour mark, it's time to prepare to prepare.

How I'd Do It Better

The solution, I believe, is in the section above. Eliminate the 96-hour and 120-hour forecasts and instead delineate a coastal *risk area* at the point that preparing to prepare should begin. The risk area would be issued 48 hours to 84 hours in advance depending on the size and strength of the storm—in other words, the level of risk at the coastline.

There are entities—the U.S. Navy, for example—that need forecasts for time frames longer than three days. Some system would have to be devised to help determine when to move the fleet, out of Norfolk, for example. Because of the long lead times involved, commanders need to know when a hurricane is just a possibility—even though the possibility is slight. The challenge is to avoid doing harm to public communications in filling this need.

Put the Forecasts Out Earlier

This is a touchy subject because it pits the scientists, who want to make the best forecast possible, against the communicators, who want to be sure the message is always correct. Currently there are three schedules that coexist in hurricane world:

- The timetable according to which worldwide weather data is collected and the computer models are run.
- The timetable under which advisories are issued.
- The time that TV newscasts start.

When a storm threatens, people turn by habit to TV at the scheduled news times. They expect to get the latest information on the storm, but most often they don't.

It's obvious that news programs that start at 4:00 P.M. or 10:00 P.M. give out advisory information that is already between two and five hours old. But inaccurate or incomplete information is actually given out much more often than that.

Two important things need to happen between the time the National Hurricane Center finishes a new advisory and it gets communicated to the public on the television news:

- Broadcasters need to digest the information and build the graphics they'll use to explain the latest data and the new forecast.
- Local emergency managers and political leaders need to prepare themselves to communicate to the community what action is being taken or needs to be taken by the government, businesses, and individuals.

If the system were organized properly, you would be able to tune into the news at prescribed times each day and *know for sure* that you were going to get both the latest on the storm and the latest status report and suggested preparation procedures from local government(s). You should be able to watch for twenty or thirty minutes—plus or minus, depending on how severe the storm is and how big the metropolitan area affected is—and get the whole picture.

If you live in Miami—Fort Lauderdale, for example, you should know that you could tune in at 5:00 P.M. and get a schedule of what storm information is coming when. For example:

5:00 New hurricane advisory and local information—local TV station
5:10 Live National Hurricane Center analysis
5:15 Miami-Dade County Emergency Management
5:30 Broward County Emergency Management
5:45 Monroe County Emergency Management

(Broward County, in South Florida, has Fort Lauderdale as its the major city. Monroe County includes the Florida Keys.) In a perfect world, the schedule would be repeated at 6:00.

In my opinion, if the communications system were organized and reliable, it would instill confidence in the public. It would appear more professional than the existing haphazard system in which a viewer never knows when or if any official is going to speak.

At present the National Hurricane Center's advisory schedule makes organizing a communications schedule as suggested above impossible. The nominal arrival time for the late-afternoon advisory package is currently 5:00 P.M. EDT. In practice, local governments normally learn what the advisory is going to say between 4:15 and 4:30. The Public Advisory and the Forecast Advisory are sent to the media between 4:30 and 4:45, although sometimes they slip to closer to 4:55. The Forecast Discussion often arrives after 5:00 for significant storms that are a threat to land.

There is not enough time for either local governments or the media to reliably and completely digest the important information that comes in the advisory and produce what they need to present during the 5:00 newscast. This is why you often see wrong numbers, incomplete graphics, and less than coherent presentations by weathercasters near the beginning of a 5:00 news program. They have not had time to properly read and understand the new information, and in some cases the full advisory package has not even come in. In fact, the graphical part of the package *almost never* comes in before the newscast starts.

How I'd Do It Better

The hurricane advisory schedule should be built around television news times. That means moving the advisory times to an hour earlier. If the advisory schedule were 4:00 A.M., 10:00 A.M. (could be 11:00 A.M.), 4:00 P.M., 10:00 P.M. EDT (with the advisories released to the media a half hour before that), the media and government would have time to present a more accurate and comprehensive picture of the storm at the main local news times of 5:00 A.M., noon, 5:00 P.M., and 11:00 P.M. Cities with news at 4:00 P.M. and 10:00 P.M. would also have the latest data, even if the whole package of information would not be available. (The problem is less acute in central-time cities, except in the late evening.)

The hurricane forecasters naturally resist this idea because they are concerned that they will be making forecasts with old or incomplete data. The longer you wait, the more data you have. Always. The data-collection system and the computer models run on a fixed schedule, a process that, in most cases, can't be speeded up.

But the question is, Are you really better off with an incrementally better forecast if it doesn't get communicated well and if a cohesive communications plan for *all* of the emergency information related to the storm can't be built around it?

I think the answer is likely no, but I would like to see the research. How much different would hurricane forecasts be if the advisories had to be issued an hour earlier? It would probably require a parallel forecasting effort by especially trained meteorologists at the National Hurricane Center. As I said, my guess is that the forecasts would not be significantly less accurate. I'm sure the communications system would work much better.

The American Media Are Often an Inaccurate Filter of Emergency Information

Have you ever been interviewed by a news reporter and then watched "your" story on TV or read it in the newspaper? I'll bet that most of the time you said to yourself, "Well, that's not quite right! They're only telling part of the story." Or maybe something stronger. Unfortunately, this is the nature of journalism. Uninformed reporters gather a set of facts, sort through them, and using their judgment and experience assemble a story *representative* of those facts, but what results is normally not a completely full or accurate portrayal of the story.

For the most part, that system works day in and day out, but it fails us in an emergency. We don't want emergency instructions filtered by uninformed, untrained communicators. It's a system that doesn't work and will never work.

Research done after every hurricane proves it. Dr. Jay Baker of Florida State University and others have conducted innumerable studies trying to understand why people did not follow emergency managers' instructions to evacuate, or why they evacuated without being told to. In all cases, a large percentage of the people interviewed said that they didn't understand what instructions applied to them or what they were supposed to do. *The American emergency communications system doesn't work.*

A significant part of the blame for the breakdown goes to the government. Its systems for organizing and delivering information are poorly coordinated and unclear. Additional blame, however, goes to the managers of television and radio outlets in the United States. They have not made it their mission to embrace their role as the communications arm of the emergency system. Any manager of a communications outlet whose viewers or listeners in large numbers misunderstand or misinterpret emergency information being broadcast is failing to fulfill a moral, if not a legal, responsibility to serve the public. We know the communications system doesn't work, but so far neither the Federal Communications Commission nor broadcasters have been motivated to fix it.

How I'd Do It Better

All television and radio stations should be required to have a Plan B, especially for emergencies, as a condition for holding a broadcast license. The major elements of my emergency broadcast system would be these:

- Except in extraordinary situations, all radio and TV stations would be part of the system.
- A triggering mechanism would be developed that would require stations to move into emergency mode.
- Stations would either be designated as an "emergency information station" or they would rebroadcast one of the designated station's signals.
- All designated emergency information stations would be required to have a set of key employees trained by local emergency management and relevant government agencies.
- Designated emergency information stations would be required to have their emergency plans and capabilities approved by local emergency management agencies and the FCC.
- The degree to which a broadcaster strengthens the physical facility, installs and maintains backup systems, and provides public service during an emergency would be noted and evaluated. This evaluation would be a significant part of the FCC's assessment of the worthiness of a broadcaster to hold a license.

A system with these key elements will assure that emergency information is being "handled" by at least some people with background and training in the subject. That should lead to the broadcasting of clearer and more accurate information. In addition, the inclusion of broadcasters in emergency planning should improve the overall quality and organization of the information coming out of the government.

Emergency communications capabilities cannot continue to be an afterthought in the American private broadcasting system.

Organize an After-the-Disaster Communications System

In the United States we have no organized plan for communicating detailed emergency information to the public *after* a major hurricane or other disaster. As a result, many people are often unaware of aid that's available to them on a timely basis, even if it's nearby. In large metropolitan areas, critical information for one suburb is often irrelevant to another part of town fifty miles away. But each suburb's critical information is not separated out and available in a way that it can be easily found by needy residents. Since the information flow is not organized, neither suburb is served well.

After the Northridge, California, earthquake in 1994, FEMA was so frustrated by its inability to get emergency information out through the local TV and radio stations that they ordered a raft of satellite dishes. People with the dishes could then receive a special satellite channel FEMA was programming.

The Northridge quake demonstrated the kind of problem that develops when a disaster dramatically affects only part of a large metropolitan area. People in the affected area are in desperate need of information long after people on the other side of the city are back at work. The communications system after Hurricane Andrew in 1992 broke down for exactly the same reason.

After Hurricane Wilma in 2005, most radio stations resumed playing music and running car commercials at least part of each hour not long after the winds died down, although more than 50 percent of the five million people in the South Florida metropolitan area were without power. In my opinion, that was grossly irresponsible. People were desperate for local, detailed information, and it was difficult or impossible to find.

How I'd Do It Better

A cooperative arrangement between local broadcasters, local emergency management offices, and the FCC would allocate radio stations to specific towns or cities. The system would work like this:

- Each defined jurisdiction—a county, city, or other geographic division—in a metropolitan area would be assigned to one or two radio stations.
- Some stations would broadcast "globally," providing general information to the whole area.
- Global stations would broadcast lists of cities and their assigned radio partners.
- Individual cities or local governments would work directly with their assigned stations to get the word out.
- At some point, when a sizable part of the area was back to normal, government entities would buy time on private radio stations to continue the emergency broadcasting services, or the public radio stations would be used to provide information to people who continue to need it.

Only with this system or something similar will a city or county government have a conduit to reach its citizens in a large, diverse metropolitan area. In addition, of course, language issues would have to be addressed using a similar allocation of stations.

Create a Television Channel Just for Evacuation and Sheltering Information

Evacuation and sheltering are the most important components of any community's emergency system. It is well known that the storm surge is the greatest threat to human life when a hurricane comes ashore. The challenge of making an evacuation system work in a complex metropolitan area is daunting. The task comes down to telling a given frightened resident in Location A how to get to a safe place, Location B. And the message has to include what transportation is available, what roads are open, what bridges are locked down, and so on.

The Internet is an excellent tool for providing this kind of information. Enter your address and the Web site could respond by giving you *your* critical information. But the Internet is not going to help people already in their car or on a bus, and not everybody has a computer or would understand how to use it. Since the information is changing all the time, the only outlets available are radio and television.

The problem is that television and radio stations that cover a wide area have too much general storm information to report as it is. There would never be enough time to provide information on a neighborhood by neighborhood or on a street by street basis. There simply is no mechanism in our current system that can solve this problem.

How I'd Do It Better

Most communities have at least one, and sometimes two or more, public radio and television stations. Most of these stations raise money by holding a telethon or telephone fund drive once or twice a year. During a hurricane emergency in a coastal area, these outlets should be dedicated to evacuation and sheltering information.

The same mechanisms that are used to raise money could be used to answer residents' questions. Imagine a telethon where the on-camera phones were tied into the local government's hurricane hotline. All general-information channels, mostly network affiliates, would tell people who need specific, local evacuation or sheltering information to tune to these public channels.

The programming might be tedious and boring, but it would be lifesaving. Close-up maps and lists of locations would be presented showing bus pickup points, bridge and road status reports, the status of shelters, and so on. A large region would be subdivided with the relevant information delivered on a schedule so residents in each part of the area would know what time in the hour they should tune in.

Providing this information in a format and via a medium that all residents can receive and understand, and act accordingly, should be among the highest priorities of local emergency management offices.

The National Weather Service Should Fix Its Graphics

There are an amazing number of errors in the graphics the National Weather Service, and occasionally the National Hurricane Center, puts on the Internet. Let me rephrase that. If you know what the graphic is seeking to communicate, you can usually figure out what it says. But if you rely on some of the graphics as your main source of information, you're going to be misled.

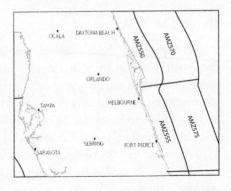

The most egregious example can be found on the home page of any National Weather Service coastal office. Take, for example, the Melbourne, Florida, office at http://www.srh.noaa.gov/mlb/. (An excellent NWS office, by the way.)

Just offshore you'll notice four outlined zones. The two immediately at the coast are the Coastal Waters Marine Zones. The boxes farther offshore outline the Offshore Waters Marine Zones.

Notice that the lower Coastal Waters Marine Zone—NWS zone AMZ555—goes from well north of Melbourne to well south of Fort Pierce. It's actually from Cocoa Beach to Jupiter Inlet. (Why they wouldn't label the endpoints of the zone is another mystery for another book!)

The problem occurs when a hurricane warning is issued from Fort Pierce north to, say, Jacksonville. The NWS will color the *whole* AM555 zone red, not just the part from Fort Pierce north. People south of Fort Pierce are *not* under a hurricane warning, when the graphic makes it look like they are.

It looks exactly like many of the maps you see on television showing hurricane watch and warning areas in different colors along the edge of the coastline. The bottom line is: it is misleading. And it is misleading at a time when getting correct information is critical.

The National Weather Service does a version of the same thing on their 3-Day and 5-Day Forecast Cone graphics. They color the coastline, but sometimes in an imprecise way. And they don't label the end points. People who live on the affected coast, however, will look hard at the minute details to see what it means to them. The graphic should be correct, and it should be clear.

How I'd Do It Better

In both cases these graphics are computer drawn, rendered, and published. No human being touches them, screens them, or corrects them. In my opinion, no graphic showing critical information should be allowed to be automatically generated by a computer if compromises in accuracy have to be made. "Almost right" or "technically right" is not good enough when you are dealing with information people might use to make lifesaving decisions.

All graphics issued by the National Weather Service, including the National Hurricane Center, should include the names of the areas affected, in this case the limits of the watches and warnings, and be made so the details are discernable and correct. In my opinion, these broad-brush graphics should be eliminated and replaced by stand-alone graphics with clear labeling that only deal with the areas that are most threatened.

Evacuation Management

Making Sure People Know That They Are in an Evacuation Zone

An evacuation is ordered for part of the coastal area. There's a wild frenzy as residents and tourists try to figure out what it means for them and whether they are directly affected. The maps shown on television don't have a lot of details or landmarks, so tourists and new residents are lost. Many people stay home because they are confused or misunderstand the instructions, endangering themselves and creating an impossible burden on emergency management after the storm.

A scenario something like this is played out in every hurricane that threatens a densely populated area. The coastal neighborhoods are too complicated and cover too much area for maps to be simple enough to be shown on television. It's rare that an evacuation plan is as simple as "Everybody east of I-95 get out." And even if it were, tourists and new residents, which can amount to thousands of people, may still be confused. If you were on vacation in Italy, would you know where a given route number is and how it pertains to you?

How I'd Do It Better

This problem needs to be attacked in two ways. First, evacuation zone boundaries need to become part of a coastal resident's everyday experience. Second; the ambiguities that exist in many evacuation communication systems need to be eliminated. Most complex jurisdictions have color-coded their evacuation areas and standardized their maps. That's a good start, but evidence shows us that it's not enough.

On the first issue, imagine a small sign at the boundary of an evacuation zone that says, for example, YOU ARE NOW ENTERING A RED EVACUATION ZONE. Over months and years, as residents drive by that sign, the boundary would become part of the local landscape. Just common knowledge. Boundaries between, say, red and orange zones would have similar signs. The point is, most residents would know whether they live and work in an evacuation zone.

Boundary signs alone, however, are not enough, especially in transient or tourist areas. In these areas a colored band could be added to each street sign designating the "color" of the evacuation zones. The message at the time the evacuation is ordered might be "If you're visiting and you're not sure if you're in a red zone, look at the nearest street sign for the colored band."

In large metropolitan areas—such as Miami, Tampa, and Long Island, New York—where partial evacuations might be ordered, another step needs to be taken. In Miami-Dade County, for example, the Red Zone includes the cities on the barrier islands, including Miami Beach, Bal Harbour, Key Biscayne, and others, but it also includes the small, residential islands in the middle of Biscayne Bay. Sometimes only the small islands need to be evacuated. On other occasions only the south part of the Orange Zone on the mainland might be at risk. Both the Red and Orange Zones need to be subdivided.

In these complex areas, more zones are needed, or the large color zones could be broken down by number. The boundary sign might read YOU ARE NOW ENTERING EVACUATION ZONE RED 2, for example. The numbers would be added to the color strips on the street signs as well. With this system, an announcement might be "An evacuation is ordered for Red Zones 1 and 2, along with Orange Zones 3 and 4." Residents and tourists would no longer be confused about whether they are included.

Currently, emergency managers rely on statements like "An evacuation is ordered for the Red Zone south of 96th Street." This doesn't deal with the reality that the words are meaningless to a significant percentage of the people who need the information, especially tourists and elderly residents who don't drive. Any emergency communications, but especially messages related to evacuations, must be clear and unambiguous. The current system in most parts of the country is far from it.

Living Successfully
in the
Hurricane Zone

6

Ready, Set, Hurricane!

What shape will you and your family be in after the next hurricane? How about your home, your business, your boat, your car, and all the other parts of your life that are in the way of the storm?

The actions you take well before there is a storm threat can make the difference between a major storm's being an aggravating inconvenience or a family catastrophe. A relatively weak hurricane can be either a minor bother or a life-changing event. You choose.

PRIOR PLANNING PROVIDES FOR PROMPT PREPARATION AND POSITIVE POST-STORM POSSIBILITIES!

Hurricanes are different than other disasters that people face. There is a substantial warning period, which means that there is an opportunity for meaningful preparation. Planning for earthquakes, for example, is a worthwhile pursuit for people living In quake-prone areas, but the aftereffects are far less predictable. in Kobe, Japan, they knew they were likely to face a significant earthquake (two big quakes had rocked the area in the mid-1940s), so the standards for construction of buildings and public infrastructure were very high. After the devastating effects of the big quake in 1995, the Japanese and other earthquake experts learned how little they knew.

In spite of the tremendous damage we've seen from hurricanes in the last few years, the potential effects of a given-strength hurricane coming from a given direction are relatively well understood . . . and therefore those *potential* disruptive effects can be diminished or eliminated.

Think of it as a battle . . . you against Mother Nature. You need tactics and a plan. This book is about making that plan, down to the small but important details. Additionally, I recommend that you take advantage of resources available on the Internet.

Even if FEMA and the federal government emergency-response system were running on all cylinders . . . which unfortunately they are not . . . there would not be enough resources to assist more than a small percentage of the people in a metropolitan area after a significant hurricane. Able-bodied people that don't prepare make it impossible for first-responders to concentrate on those folks who, through age, disability, or bad fortune, are unable to handle the storm or its aftermath without help.

The bottom line? If you live within about seventy-five miles of the Gulf of Mexico or the Atlantic Ocean, you have three options:

1. Make preparing for a hurricane and defending yourself, your family, your property, and your business an integral part of your life.
2. Roll the dice and hope that the odds go your way.
3. Move.

Unfortunately, most people choose door number two . . . endangering themselves, their families, and the people that come to help. My hope is that this book and the accompanying Web sites will provide the help, motivation, and understanding so more people plan and prepare. There is no other option for living successfully in the hurricane zone.

Web Sites for Hurricane Planning

www.HurricaneAlmanac.com—Updates to this book, hard-to-find hurricane supplies, links to other Web sites.

www.OneStorm.org—*Free* online hurricane planning. The Web's most comprehensive guide to preparation, evacuation, and recovery issues. A joint project of Bryan Norcross and OneStorm.

Making a Good Family Hurricane Plan

A good hurricane plan is like personal insurance. Invest a little time and energy now and you'll get paid back in spades if a hurricane happens.

Your plan will be different depending on the location of your home and whether it's a house or apartment. Also, obviously, the makeup of your family will play into your planning. There are physical things—shopping, home protection, etc.—you are going to have to do, but you can't get started without sitting down, gathering some information, and making decisions.

The Big Questions

Are we going to have to evacuate?

How are we going to take care of ourselves during and after the storm?

How are we going to protect our stuff—the house, the contents, the boat, everything?

Each question deserves thought and planning, so I've devoted a separate section to each one to help you sort through the issues. Also, depending on your answer to Question 1 and where you live, Question 2 takes on different dimensions.

A Family Project

Experience has shown that hurricane plans that involve *every* member of the family are the most successful. You can't hide the fact that something big is happening, so give everybody a role in making the best of the situation.

Make the kids responsible for getting the games together that you are going to play in the car, for example. They'll feel good that their work made the trip less tedious.

Hurricanes can be traumatizing experiences, but the worst aspect is the unknown. Involve the whole family, talk about the plan for being safe, and the experience may be less traumatic.

Family Communications Plan

IF YOU DON'T DO ANYTHING ELSE, DO THIS!

Think about all the people in anguish after Hurricane Katrina because they didn't know the status of other members of their family. You can avoid any chance of this happening to you by taking a couple of easy steps now.

Doing the Minimum

- Contact a friend or relative *out of town* and ask him or her to be your family's emergency contact.
- Before the storm, be sure that *every* member of the family has a piece of paper on them that says, for example:

EMERGENCY CONTACT
AUNT MILLY IN NJ
201-555-5555

- Call Aunt Milly before the wind starts blowing to tell her exactly where you are and what you are planning to do.
- Be sure everybody knows that they should call Aunt Milly if they get lost or anything bad happens.

It's important that your main contact person is out of town, because local calls are more likely to be disrupted after a storm. Both ends of local connections are subject to problems.

Doing More

If you want to do this right, take a few more steps. Here are some ideas:

- Print out business-size cards with your family information as well as your out-of-town contact(s) information. Have them laminated in plastic. (Kinko's can usually do that.) Be sure everyone in the family has one.
- Set up more than one out-of-town contact.
- Be sure everybody knows how to text-message on a cell phone. While mobile calls often won't go through after a hurricane because the system is overloaded, text messages often will.
- Set up your family's cell phone address books with some key e-mail addresses. Most cellphones can send e-mail to a computer as well as text messages to another cell phone. You just have to know how to do it.
- Learn how to send e-mail to cell phones. All mobile-phone companies have a way to send e-mail from a computer directly to cell phones. Or you can send a message to any cell phone by e-mailing to XXX-XXX-XXXX@teleflip.com. The message will be forwarded to any cell phone provider in the U.S.

Planning for Drinking Water

Your most important hurricane supply is water. You can survive without food, but in the post-hurricane heat you're going to need plenty of water. Here are the guidelines:

Five gallons per person. Plan on having five gallons of fresh water per person in your household on hand. That supply will last about a week considering you'll likely have some juices and other drinks.

Some bottled, some tap. You don't need to buy bottled water to meet all your needs. In fact, it will be easier if you don't. Buy a couple of gallons per person, and fill plastic bags or collapsible water jugs with tap water for the rest.

Freeze bags of water. Fill your collapsible water jugs or zip-bags about 90 percent full and put them in the top of the freezer. Move the food to the bottom. Fill as many nooks and crannies as you can. The less air you have, the longer the freezer will stay below 32°F. These frozen bags will likely start to melt after about two days without electricity, so use your other water up first.

Freeze Water Early

It will take a while to freeze all that water. Plan to stock your freezer with water the day before you anticipate that a hurricane watch might be issued.

Water jugs. If you are using tap water instead of store-bought bottled water, old soda bottles are stronger than old milk jugs, which can leak over time. Collapsible water containers, made for camping, or large jugs from the home-improvement store are the best.

Sanitize old jugs. To sanitize old bottles or jugs, put a teaspoon of nonscented chlorine bleach in about a quarter-full jug of water. Slosh it around good and empty it. Rinse out the bottle or jug, then you can fill it with water for drinking.

Water heater. If the power is out and the water supply is not drinkable, the tank of your water heater is holding good water. Turn off the power and the intake water valve before using it to keep contaminated city water from flowing in.

Evacuation Decision Making

Primary question: Do I live in an evacuation zone?

Unfortunately, the answer is not always easy to come by. Governments in many locations have done precious little to make it readily apparent. You shouldn't *have* to work to find out, but sometimes you do. Look in the phone book, call your emergency management office, check your emergency management office's Web site, or call your city hall. You'll get one of three possible answers:

(NO)   You will never be ordered to evacuate.

(YES)   You will almost *always* be ordered to evacuate for a hurricane approaching from the water, and you may be ordered out for storms coming from other directions.

(YES)   You will only be ordered to evacuate for *strong* storms approaching from the water. There's very little chance you would be ordered out for storms coming from other directions.

If the answer if no, you can move on to preparing your home.
If the answer is either version of yes, read on.

Why evacuate?

Evacuations are ordered for four reasons:

- There is a risk that the storm surge will sweep over the land, destroying buildings and drowning people caught at ground level.
- There is a risk of people being stranded on an island, in a high-rise, or in other inaccessible locations.
- People are living in housing not suitable for hurricane-force winds (like mobile homes).
- Some high-rise buildings require people to leave during hurricanes. Elevators are locked at the top, air-conditioning is shut off, and other systems are shut down to protect them from storm damage.

We'll deal with each of these issues separately.

Storm Surge Kills

If there is *any* risk that the storm surge might reach you and your family, *you must get out*. Even six inches of fast-moving water will sweep you off your feet and wash a car off the road. Your local emergency management office knows how high the storm surge may reach in your area. The numbers vary tremendously, from about 10 feet to 30 or 40 feet, depending on your location on the coast. (See "Storm Surge," page 117.)

DO NOT TRY TO GUESS WHAT THE STORM SURGE WILL BE! GET THE FACTS.

The computer programs that forecast surges are very accurate. If emergency management says that your area could be inundated, you have no options. You must leave.

Stranded by the Surge

In some locations, like the southeast coast of Florida, many people in the evacuation zone fall into this category. Well-built high-rises on barrier islands are expected to survive any storm surge. People who protect themselves on a low floor above the expected water level are likely to ride out the storm successfully. However, the aftermath may be difficult, and potentially life-threatening. Among the problems that may be encountered are:

- Cars left at street level or in parking garages on lower floors will likely be ruined.
- Streets near the ocean, and perhaps a good way inland, will be covered with sand and debris, so emergency vehicles may not be able to respond.
- The power, water, and phone service will likely be out.
- The causeway or access road to the mainland may be cut off for some time.
- There may be no practical way for the government to provide security.
- You may have to fend for yourself for an extended period of time in a very unpleasant environment. You may well be stuck.

The alternative is to go to a safe spot away from the water. Let's not kid ourselves, unless you have wonderful friends with a big, strong house well inland, this isn't going to be a picnic no matter what you do. But it will be infinitely better if you have a plan.

Vulnerable Housing

Mobile and most manufactured homes are evacuated during hurricanes. While some are stronger than others, there is no safe spot in the center of most mobile homes. All mobile home residents need to make a plan to leave.

Many—even most—houses in the hurricane zone are not as strong as they should be. It is a risk staying in any house that is not shuttered with the roof and doors checked and secure. The risk is significantly lower, however, if you keep your family in the center of the house in a properly fashioned safe spot (see page 320). Even the exterior walls in most homes are vulnerable to penetration by flying debris in an intense hurricane.

If you are uncertain of the strength of your house, have an inspector give you an evaluation. Your house may be secure for category 1 or 2 hurricanes, but be vulnerable in stronger storms. This is not a decision to make when the storm is coming. Evaluate your house now and make your plan.

High-Rises Evacuated

Don't get angry at your high-rise management if they make it uncomfortable for you to stay in your building. They are thinking about what it will cost them after the storm if critical systems are damaged because they weren't secured. Talk to your building management so you can accurately assess what staying at home will be like—if it's permitted at all. And remember, even if you stay in your building, you should never stay on a high floor during a period of high winds. See "Apartment/Condo Preparation Checklist," page 317.

Outta Here

Many people, unfortunately, have a two-word evacuation plan, "Outta here!" But, of course, that's not a plan. Wealthy people figure they'll take a plane somewhere. That's not a plan either.

You can go *anywhere* if you commit yourself to doing it early in the process—before a hurricane watch is issued. But most people end up not leaving early, and they end up trapped and planless. Roads get clogged and airplanes get full or stop flying *before it is certain that the worst of the storm is coming*. Get over it. Make a realistic plan and stick with it.

Your Evacuation Plan

The decision to evacuate is always a hard one. Once it's made, you've got to get down to the nitty-gritty and get it figured out. The questions come down to:

- *When will we leave home?*
- *Where will we go?*

- *How will we get there?*
- *What will we take with us?*

It's time to make a plan for your evacuation.

Planning for Evacuation

Knowing there's a good family plan for being safe during the next hurricane should give everyone peace of mind. Once the family decision to evacuate has been made, it's time to go through the questions one by one.

- *When will we leave home?*

The easy answer is: As early as possible. When you leave is, to some degree, controlled by where you are going. If it's a short trip, you can wait awhile. If, however, you need to drive a considerable distance, you may want to leave even before the evacuation is ordered. A common mind-set is "I'll wait a while and see how it goes." This is equivalent to saying, "I'm willing to take a chance with the lives of my family." If you don't go early, there's a chance you won't be able to go at all. The roads may become too congested, or the storm may speed up. You could end up stuck.

Go. Get on with it. Be happy if nothing happens.

- *Where will we go?*

For most people, this is the most difficult issue. If you're lucky, you can go to a friend's house that's relatively close by or to an inland hotel. You are looking for a place that is safe—is well built with hurricane protection—where you can stay for up to a week. (Hopefully not that long, but plan for it.)

Whether it's near or far, know where you're going. Don't just head out hoping to get a motel down the road somewhere. The odds are you won't find one. Your goal should be to stay as close to home as possible in a safe structure. In general, especially if you live in a house, the sooner you can get back, the better your post-hurricane life will be.

If you have no place to go, there are, of course, public shelters. They should be your last resort, but they will be safe. Finding a shelter, especially in a large metropolitan area, can be difficult. (Governments have been notoriously poor at setting up a communications system for this purpose.) Call your emergency management hotline if you don't know where to go.

Pets will play into your decision as well. Most shelters won't take pets. It's critical that you plan in advance what you are going to do if you have to leave home. See "Your Pet's Hurricane Plan," page 268.

- *How will we get there?*

Most people will evacuate in their car. It's the best way to go because you can take important things with you, and you get your car out of danger. Remember, however, that your car is vulnerable to being damaged if you are not careful where you park it. Also, be conscious of the security of your parking place if you leave valuable items in the trunk. See "Where to Park Your Car," page 274.

Plan your route before you leave. Internet map programs can be a big help. You often get a different route on Yahoo! than on Google, so print out maps from more than one service for use in case the roads on the route you have chosen are clogged.

If you don't have a car, buses are normally available to take you to a shelter. Finding a bus pickup point may be difficult in large metropolitan areas (again governments in many areas have not established a workable system). Your only solution might be to call your emergency management hotline as soon as it is open to find out where the busses stop near you. Do *not* let a bad communications system keep you from getting on the bus. Call the police nonemergency number as a last resort if you can't figure out what to do.

- *What will we take with us?*

At the back of the book is an "Evacuating Checklist" (see page 307) to help you pack the car. Think carefully and double-check. You won't be able to go back for something you forgot.

If you are going to a shelter, you are very limited in what you can take with you, but you will be much more comfortable if you bring the right things. There's also a "Shelter Checklist" at the back of the book (see page 313).

If you evacuate and the hurricane doesn't come, it's okay to get annoyed. But there is no alternative. *Not* evacuating is to play Russian roulette. Make the decisions when you are thinking clearly, not under the stress of an approaching hurricane.

Planning for Staying in a House

Be sure you read "Evacuation Decision Making" (page 207) before proceeding here. This section assumes that you have carefully thought out your choices and decided it will be safe to stay at home.

Staying at home is your best option if can do it, but you have to prepare. There are three stages to preparing a house:

Long-term projects. Protecting the windows and doors, evaluating and strengthening the structure, and taking steps to make your preparation process easy when a storm threatens.

Hurricane season projects. Reevaluating the house and its hurricane protection status. Repairing or supplementing the protection systems in place. Organizing tools, shutters, and anything else that you will need if a hurricane threatens.

Quick action before the storm. Taking the final steps necessary to secure the house so that it provides a safe haven for your family.

Each step requires thought and planning. If houses had been built and were being built with hurricanes and safety in mind, this process would be easy. Unfortunately, in most of the hurricane zone, storm readiness does not come as standard equipment. Therefore, homeowners must take preventative action.

Long-Term Projects

The most vulnerable parts of most houses are the windows, the doors, and the roof. The techniques used in building the house can also dramatically affect the strength of the structure. See "Protecting Your Home," page 218.

Hurricane Season Projects

The job of protecting your home never stops. *Every year* you have to double-check that you have everything you need so you can take action quickly when a hurricane threatens. Beginning on page 325, you'll find the shopping lists you need to get yourself ready so the process of final preparation is as easy as possible. Investing a few hours early in the season is *guaranteed* to save you days, weeks, or months of aggravation if a storm hits.

Also, run down the "House Preparation Checklist" on page 314, and the "Staying-in-a-House Checklist" on page 320 to be sure you have what you need to implement your plan on short notice.

Quick Action Before the Storm

When a hurricane watch is issued—if not just before—it's time to spring into action. If you have done your job properly earlier in the season, you'll only need to top off your supplies and put your family preparation plan into effect. See page 282, "During a Hurricane Watch," as a guide to make sure you have taken all of the important steps.

Since you are staying home, you *must* create a safe spot for your family to ride out the storm away from the outside walls, windows, and doors. See "Staying-in-a-House Checklist," page 320.

Planning for Staying in an Apartment/Condo

Be sure you read "Evacuation Decision Making" (page 207) before proceeding here. This section assumes that you have carefully thought out your choices and decided it will be safe to stay in your apartment or condo.

If you live in a high-rise, be sure to check your building's rules. The managements of some high-rises require tenants to leave, or make it very uncomfortable if they stay.

Staying at home is your best option if can do it, but you have to prepare. There are three stages to preparing an apartment or condo:

Long-term projects. Protecting the windows and doors. You must have shutters or other hurricane protection if you intend to stay in your apartment during a hurricane.

Hurricane season projects. Reevaluating the hurricane protection systems. Be sure the shutters will close and lock, if you have them. Double-check the plans of the building management so you know what to expect. Be sure you have what you need to take quick action if a hurricane threatens.

Quick action before the storm. Taking the final steps necessary to secure the apartment or condo so that it provides a safe haven for your family.

Each step requires thought and planning. If apartment buildings were planned right, there should be a safe place on a lower floor designed to serve as a hurricane shelter. Unfortunately, in most of the hurricane zone, storm-readiness is an afterthought, if it's a thought at all. Therefore, each resident has to take action to secure his or her apartment or condo and find a safe place to ride out the storm.

Long-Term Projects

The most vulnerable parts of apartments and condos are usually the windows and doors (including sliding glass doors). The wind forces increase above the ground, so shutters or other protection systems are essential for apartments and condos—especially in high-rises. See "Protecting Your Home," page 218.

You may also have vulnerabilities due to the building's roof or your neighbors' lack of hurricane protection. Talk to your building management to be sure you understand the risk that something out of your immediate control will cause a problem.

Hurricane Season Projects

Every year you have to double-check that you have everything you need so you can take action quickly when a hurricane threatens. Beginning on page 325, you'll find the shopping lists you need to get yourself ready so the process of final preparation is as easy as possible. Investing a few hours early in the season is *guaranteed* to save you days, weeks, or months of aggravation if a storm hits.

Run down the "Apartment/Condo Preparation Checklist" on page 317, and the "Staying-in-an-Apartment/Condo Checklist" on page 322 to be sure you have what you need to implement your plan on short notice.

Quick Action Before the Storm

When a hurricane watch is issued—if not just before—it's time to spring into action. If you have done your job properly earlier in the season, you'll only need to top off your supplies and put your family preparation plan into effect. See "During a Hurricane Watch," page 282, as a guide to be sure you have taken all of the important steps.

Since you are staying home, you *must* create a safe spot for your family to ride out the storm away from the outside walls, windows, and doors. See "Staying-in-an-Apartment/Condo Checklist," page 322.

Protecting Your Property

Protecting people is easier than protecting property because people can move out of the way of a hurricane, or at least take cover in a safe spot. Most of the things you own, however, are at some level of risk unless you take some action.

- **Protecting your home, apartment, or condo** comes down to shutters and other hurricane protection systems. See page 218 for a guide to the options you may have.
- **Protecting your personal property** means having a place to put valuable, irreplaceable, or sentimental things where they will be safe from the wind and the water.

The best plan is to protect your valuables by stowing them above first-floor level in an interior part of your home or any other location that is secure and will remain dry. See page 286 and go over the process so you're ready for a hurricane warning.

- **Protecting your car** means thinking carefully when you park it before the storm. There are no guarantees, but you'll find some ideas on page 274.
- **Protecting your boat** depends on its size and where you normally keep it. Boats are a difficult challenge, but see page 271 for some of the things that you can do.

Plastic storage boxes, a fireproof and waterproof lockbox, heavy-duty contractor bags, and duct tape will be your best friends if water gets into your house.

Protecting Your Home

Making your home hurricane-resistant means protecting and strengthening the most vulnerable parts. In most cases, the areas of concern come down to the windows, the doors, the roof, and the "connections" that hold the house together. Not all houses can, as a practical matter, be reinforced to be "major-hurricane-proof," but paying attention to these components will make a significant difference in the strength of the structure.

To be effective, hurricane protection must deal with two threats:

- The *pressure* of the wind pushing and pulling on your house.
- The *debris* that gets caught in the wind stream smashing into your structure.

Don't Tape Your Windows

Taping windows is a waste of time and money, and it's dangerous and messy. Tape does not strengthen the glass and could give you a false sense of security. Flying debris will smash a taped window and injure whoever is behind it as if the tape weren't there. Also, if you're lucky and the storm doesn't come, you'll have a gooey mess on the glass.

Understanding Wind Pressure

Every part of your home that comes into contact with the outside air is subject to extreme wind pressure in a hurricane, including the windows behind the shutters and the "back" of the house away from the wind. In fact, the strongest forces occur in areas where you might not expect them.

When the wind blows on the front of a house, it forces the air to

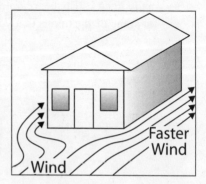

The wind speed increases on the sides of the house as the air is compressed. This creates a strong suction on the back corners.

squeeze around the sides. The high-speed wind pulling away from the back of the house creates a strong suction on the back corners. This suction (called negative pressure) can be stronger than the direct pressure from the wind on the front side. Similarly, when wind is forced up and over and around the roof, negative pressure creates an upward suction there as well. The important point is to re-

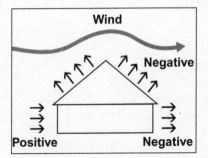

When the house is sealed, the wind passes over and around it.

member that every part of the house is subject to these strong forces. Do not try to outthink the wind and protect only one side of the house and not the other.

The Open-a-Window Myth

Standard operating procedure for years was to open a window on the downwind side of the house to "relieve the pressure" during a hurricane. That's a BAD IDEA. Air will be pulled out through the open window creating lower pressure inside the house and a bigger pressure contrast across the front wall.

Keep all the windows closed while the strong winds are blowing.

The Danger of Flying Debris

Shutters and other protection systems are designed to protect vulnerable openings from flying roofing materials, tree limbs, and other debris. The goal is to *keep the wind outside the house.*

As I discussed above, the wind creates a suction that pulls upward on the roof and outward on the downwind sides of the structure. (Depending on the slope of the roof, it can also create negative pressure on the roof on the windward side as well.)

If a window breaks, the inside of the house gets pressurized, and

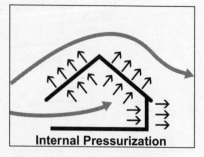

Internal Pressurization

When the wind gets in the house, upward pressure adds to the negative pressure from outside.

the upward forces on the roof are now coming from above *and* below. This pressure is what causes roofs to fail when windows break.

Shutters and other protection systems are designed to *solve one problem* only: They stop debris. They do *nothing* to stop the pressure of the wind on the windows and doors. A weak window or door can still fail even behind a good shutter.

How to Keep the Wind Outside

- The cost and strength of hurricane-protection devices varies tremendously over the hurricane zone. In some areas shutters and other devices and techniques can be selected, bought, and installed by the homeowner. It's up to the buyer to decide how much hurricane resistance he or she wants to buy. In other places, like Miami-Dade and Broward counties in Florida, only the strongest, highest-quality products are allowed to be sold and installed.

- There are two approaches to strengthening and protecting windows: shuttering standard windows or installing "impact" windows that will protect against airborne debris without shutters. See information about shutters and your other options on page 221.

- Each type of door has to be treated separately. Like windows, doors can be shuttered or replaced with extrastrong models that are designed to withstand a hurricane's winds and debris without additional protection. Many of the options are discussed on page 238. You must keep two doors shutter-free so you can get out quickly if there is a fire or other emergency.

- Roofs are extremely complicated, and you will likely need an engineer or roofing expert to get the full picture. There are, however, some things that you can check on and improve. Gable ends, the triangular shaped part of the house just under the roof can be vulnerable. Go to page 244 for more on roofs and gable ends.

The Miami-Dade Standard

Miami-Dade County, Florida, has developed a hurricane-protection standard that has been internationally adopted by the industry and other governments. No matter where you live, if the product you're considering has—these exact words—"Miami-Dade County Notice of Acceptance," you can be certain that it is a very strong system if it's installed according to the specifications.

Some products will claim to have "passed the Miami-Dade large impact test" or something similar. That doesn't mean they are necessarily bad products, and they may be fine for your installation, but they have not passed the rigorous standards to get final *product approval*.

Go to www.hurricanealmanac.com for a link where you can check to see which products have been certified.

- A stack of plywood is easily blown apart by strong winds, but nail the sheets together and it's not as likely. That's obvious to most people, but not always to contractors. Every house in the hurricane zone should be tied together from the roof to the ground. Again, you will probably need an inspector to give you the full story on how your house stacks up. See page 246 for a list of products and techniques to make houses better withstand hurricanes.

Window Protection

The first thing to decide is whether you want to shutter your existing windows or install new "impact" windows that don't require separate shutters. The "glass" in these windows contains a polycarbonate

(essentially bulletproof plastic) that is spectacularly strong and virtually unbreakable. In most cases, either option will give you good protection, although adding shutters to weak, old windows will not increase
their strength against the pressure of the wind. Also, not all of the options will work on all structures.

Plywood Shutters and Window Film

You may be thinking, "I'll use plywood." That's an option, but
not necessarily the best one, and in some areas it's not technically
legal. See page 233 before you go down that road.

Window film has its place as hurricane protection, but is not
a substitute for shutters. See page 237 for more.

There are many different kinds of shutters with many options, but
in general, here are the pros and cons.

	PROS	CONS
Shutters	Generally less expensive than impact windows.	You have to put them up or close them for each storm.
	If some of the shutters get damaged in the hurricane, they are easy to replace.	Some shutter systems are unattractive and can ruin the lines of your home.
	There tends to be less water leakage around shuttered windows.	Most shutters require maintenance.
Impact Windows	You don't have to do anything. They are there all the time.	They are relatively expensive.
	Windows seal well, keeping the water out.	After a big storm, the outer glass might be broken or the frame dented or bent,

PROS	CONS
	requiring an expensive and messy window replacement.
No maintenance is required. If locked properly, they provide extra security.	

Making Your Choice

You probably can't decide which is right for you just by looking over that list. There are other important issues to consider. For example: What kind of structure are you protecting? Wood or concrete? High-rise or low-rise? Will you have easy access to the windows or will you need a ladder? Are there some windows where you have no access at all?

Restrictions and Required Permits

Before you make a final choice be sure you know whether there are any restrictions on what you can install. High-rises, for example, often dictate the style of shutters that are permissible. And in some jurisdictions you must get a permit to install shutters on your home.

Getting a permit is a GOOD thing! The city inspector is on your side and, theoretically, will make sure the shutters are installed properly and will perform as you expect in a hurricane.

On the following pages are many of the available options for protecting your windows. This should help you make a good choice.

Shutters for a Concrete-Block Structure

Shutters come in a variety of shapes and price ranges. In general, you pay for convenience. The easier the shutters are to close, the more expensive they are, although there are some exceptions. Each major style of shutter is listed below with its pros, cons, and, in some cases, options.

Storm Panels

Storm panels come in a variety of thicknesses and strengths. Thicker is, in general, stronger in aluminum panels, but the best gauge of strength is the Miami-Dade County Notice of Acceptance. Panels without that stamp of approval can still provide some protection, but they are generally not as strong as the approved products. For example, aluminum shutters thinner than .050 will not pass the Miami-Dade tests, but they are for sale in many areas.

(PRO) Storm panels are the cheapest manufactured shutter system.

(PRO) Aluminum storm panels are relatively light; most people can handle them.

(PRO) If a panel gets dented or lost, it can be cheaply and easily replaced.

(PRO) Panels can be used to protect sliding doors, even in high-rises, with the appropriate options. (See page 242.)

(CON) Storm panels need a storage area.

(CON) Installation can be tedious and tiring.

(CON) Installation has to occur well in advance of the storm because handling panels in the wind is dangerous.

(CON) Most panels have sharp edges. You have to wear gloves and be careful.

(CON) Storm panels are generally *not* appropriate for the second floor and above. Handling panels on a ladder can be dangerous.

(CON) Because panels are tedious to take down and put back up home-owners may leave them up. Houses tend to be dark inside for weeks on end.

(CON) There are lots of small parts that can rust and get lost.

There are options to consider that make storm panels more user friendly.

(OPTION) Panels are much easier to install quickly, and require much less effort, with optional headers and sills. Secure these tracks above and below the window and panel installation becomes a relative snap.

(OPTION) Buy the more expensive aluminum panels. They are relatively light, and essentially as strong as steel panels.

(OPTION) There are a variety of systems available for securing the shut-ters, depending on the type of track and sill you have. Some clip systems make installation faster.

(OPTION) Add clear, polycarbonate panels to the middle of every window to let light inside. The clear panels can replace metal panels en-tirely on some windows, but they won't work where the shutters are close to the glass because the polycarbonate flexes.

(COST) $5–$10 per square foot.

Anchoring the Shutters to the Building

The system you use to secure your shutters to your house is as important as the shutters themselves. If you are doing a Miami-Dade County—approved installation, the type of anchor will be specified on the paperwork that comes with the shutters. In order to get a Miami-Dade Notice of Acceptance the manufacturer tested the shutters with those specific anchors. *Every* approved shutter system has approved anchors that go with it.

If you are installing the shutters yourself, check with your local building department to be sure you are following the appropriate regulations. Before choosing your anchoring system, get the shutter company's recommendation. If that's not possible, see the anchoring guidelines below. Do your homework on anchors. In the hurricanes in recent years there have been many examples of shutters blowing off of houses because they were improperly anchored.

Anchor Types

There are three types of anchors: temporary anchors you remove from the house, permanent anchors that have a screw protruding from the house, and permanent anchors that are inset into the wall with a removable screw. Here are some guidelines:

Temporary anchors. Concrete screws and especially concrete nails are *not* recommended. The holes grow larger with every use, so the connection gets weaker with time.

Permanent anchors. The PanelMate® or Tapcon®SG products are good for permanently mounting tracks and sills. But, without something to cover them up, they leave screws protruding from your house all year round.

Permanent inset anchors. The Tampin® brand machine screw anchors or various adhesive anchors are heavy-duty versions of the system you might use to hang a picture on the wall. An expanding sleeve stays in the wall, but you can put the screw in and remove it without weakening the system. Look for "sidewalk" bolts—bolts with a big, flat head— to fill the holes. You can also paint the heads of the bolts to match the house, but be sure they will unscrew easily when it comes time to put up the shutters.

Accordion Shutters

Accordion shutters fold back on one or both sides of the window. You close and lock them like sliding doors right before the storm. During Hurricane Andrew, many accordion shutters failed, the "slats" came out of the track or the shutters came off the buildings. Now accordion systems are made much stronger. Check for the Miami-Dade County Notice of Acceptance to be sure you're getting the strong version of the shutters.

(PRO) Accordions don't have to be put up. They are always there and easy to close.

(PRO) Accordions can be used on high-rises and second floors. Certain models can be closed from the inside by using a pole with a hook to pull the shutters closed.

(CON) Accordions are significantly more expensive than storm panels.

(CON) As a practical matter, you'll need a professional to install accordions.

(CON) The bulky, less-than-appealing shutters are on your house year-round.

(CON)   Accordion shutters require lubrication and maintenance.

(OPTION) Accordions are available with clear panels to let in light.

(OPTION) Some models can be locked with a key for added security.

(OPTION) They are available in a variety of colors.

(COST) $15–$25 per square foot.

Roll-Down Shutters

(PRO)   Like accordions, roll-downs are always there and ready to use.

(PRO)   One person can close roll-downs quickly.

(PRO)   When they are closed, they are an excellent barrier against burglary.

(CON)   Roll-downs are the most expensive type of shutters.

(CON)   Electric-only shutters need a backup power system so they can be opened if the power goes out.

(CON)   Most people think they are *less unattractive* than accordions, but they are still bulky on the outside of some buildings.

(OPTION) Manual-crank shutters work without power. Some electric roll-downs come with backup cranks.

(OPTION) Some models come with clear panels to let in light.

OPTION They can be built into the wall if they are installed during construction.

COST $27–$45 per square foot.

Beware of Sun Shades

Sun shades look very much like roll-down hurricane shutters but are not intended to stop airborne debris. In fact, they must be raised in high winds. Be sure you are getting roll-down shutters, and not the much lighter-weight product.

Bahama Shutters

If you have windows that you normally want shaded, Bahama shutters are a good option. They permanently attach to the wall and are propped open at the bottom so some light gets through. They are good for bedrooms and on the south side of buildings.

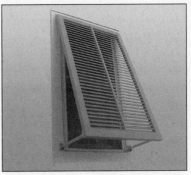

PRO They are always attached and easy to lock down.

PRO They provide shade and privacy.

PRO They are attractive in the right situations.

CON They can't be used on all windows because they block the light.

COST $25–$40 per square foot.

Beware of Old-Style Bahamas

Before the South Florida building code was changed after Hurricane Andrew, Bahama shutters were generally made with very weak slats. That is no longer the case. Be sure to get Bahamas with a Miami-Dade County Notice of Acceptance.

Colonial Shutters

On houses where colonial shutters accent the design, these are a great option. They sit on either side of your windows and easily swing or slide into place.

(PRO) They are attractive on a compatible house.

(PRO) They are easy to close and lock before a storm.

(PRO) They can be closed from the inside if necessary.

(CON) Some models require you to install an extra bar to add strength, but this is not a significant problem.

(COST) $25–$40 per square foot.

Flexible Screens

This relatively new system won't work for every opening, but in difficult-to-shutter situations it can be very effective. This extremely strong fabriclike screen is mounted well away from the opening it is meant to protect. Anchors are placed in the wall and on the ground or floor so the screen makes a tent well away from the structure.

(PRO) It will cover bay windows, garage doors, condominium balconies, and other situations that are difficult to protect with standard shutters.

(PRO) The screening lets some air through but blocks most of the wind pressure.

(PRO) Some light comes through the screening.

(PRO) This is the lightest shutter system available.

(PRO) Protection for entry doors can be installed right before the storm arrives.

(CON) The folded screening has to be stored somewhere.

(CON) Permanent attachment points on the house are always visible.

(CON) The system only works when it can be mounted well away from the window or door being protected due to the flexibility of the screen.

(CON) For big openings, more than one person is required to hang the screening.

(CON) $16–$25 per square foot. Remember, the screening has to be made considerably bigger than the opening it is protecting.

A New Version

A modification on the flexible screening system discussed above became available in 2006. This product is draped over your house or mobile home and then anchored to the ground on both sides. The "net" stops debris from hitting the windows *and* it increases the strength of the roof. This system won't work for every situation, but it is worth considering, especially for buildings that need the windows, doors, and the roof strengthened or protected.

Unapproved Version

There is a low-priced flexible-screening system that mounts flat on the house. It does not have a Miami-Dade County Notice of Acceptance because, under some circumstances, it doesn't hang far enough from the window. Flying debris will deflect the fabric and could break the glass. For deep-set windows, doors, or on patios this may be a good option, however.

Shuttering a Wood-Frame Building

Putting shutters on a wood-frame building is trickier because there are more variables. Before you start, review these considerations:

Where Will You Mount Your Shutters?

Often the windows on wood buildings are *not* inset, making shutters difficult to mount. Also, studs can be hard to find. You *must* attach your shutters to structural components, not the siding. There may be some creative things you can do, and maybe a local shutter company has some ideas, but here are some general guidelines.

If the windows are *not* inset (they protrude from the house). You'll need a shutter system that can attach to the window frame. This is not an ideal situation because all of the force of the wind is transferred to the frame. You don't know how well the window was installed, or how well it's attached to the house. But there's not much else you can do.

Storm panels, plywood, and the fabric screening system described in the previous pages are possibilities.

If the windows are inset. You have to determine whether you can attach anything to the house. If you have aluminum siding, it's likely you can't. In this case, your best option is likely to be plywood inset into the window opening.

If you can securely attach shutters to the side of your house you have more options. Storm panels with tracks and sills, for example might be a good choice. Just be sure the anchors are secure in the studs and not just in the siding.

Anchoring the Shutters to the House

A connector product called Tapcon®SG is made for this kind of mounting. One end of the connector has a long wood screw that goes into the house, the other end is threaded to hold the shutter. The down side is that the screw protrudes from the house year-round. You might consider leaving the tracks and sills attached or get rubber caps for the ends of the screws.

Plywood Shutters

Plywood shutters are rarely the best option (though on some wood houses they might be), but for some people they are the only option. Poorly installed plywood panels can be extremely harmful, however. They can fly off and severely damage property downwind.

There are two ways to install plywood shutters. The first technique, the *inset method*, is recommended if your house can accommodate it. The second way, *the overlap method*, is much harder and should only be used if you absolutely cannot, for whatever reason, get storm panels.

Plywood and the Code

In many areas plywood shutters are technically illegal. It's not so much that they can't be strong—although in general they can't pass the Miami-Dade County tests—but that installing them properly is not easy, and installing them improperly can be dangerous for you and your downwind neighbors.

The Inset Method—Recommended

Shutters installed this way performed well in Hurricane Andrew, a good recommendation. Also, the shutters are easier to make than in the second method, and the installation is often stronger. To use this method *your windows must be inset at least two inches from the exterior wall.* These shutters will sit *inside* the window opening, not overlap the wall outside. You need room to secure them with barrel bolts inside the inset, but away from the window glass.

Here are the steps. Do them *before* the storm is bearing down.

1. Measure the inside dimensions of your window openings. Buy exterior-grade plywood at least ⅝" thick.
2. Buy sliding barrel bolts at least 3" long. (For a count, see step 4.)
3. Cut the plywood to fit each window. You want a good fit so the shutter can't move around.
4. Screw the barrel bolts every 18 to 24 inches around the perimeter of the plywood.
5. Put the shutter in the inset, but not against the window frame. Slide each bolt hard against the house so it leaves a mark.
6. Remove the shutter and drill a hole to accept the slide bolt at each mark. Don't make the hole much larger than the bolt. Plan on tapping the bolts in place with a hammer for a tight fit.
7. Before you get the shutters mixed up, label the top and front of each one and the window name.
8. For large openings, you can use a piano hinge to join two pieces of plywood.

See www.HurricaneAlmanac.com for links to helpful Web sites.

Clip Systems

There are other systems on the market to hold plywood in window openings, including some spring clips. I'm not crazy about them. The ones I've seen are plenty strong, but they rust and, in my opinion, are harder to use than the slide bolts.

The Overlap Method—As a Last Resort

This method is hard work and difficult to do right, but sometimes it's the only option. On the surface it sounds simple. But when you try holding the heavy panels in place to get the holes lined up, you'll find out you need a family of grown boys. Otherwise it's a huge chore.

1. Measure your window openings and add 4" on each side. Buy exterior-grade plywood at least ⅝" thick.
2. Buy appropriate anchors (see the options on page 226). You'll need an anchor every twelve inches around the boards.
3. Cut the plywood to fit each window. Precision doesn't count here, but be sure you have at least 4" overlap on each side of the window.
4. Drill holes in the top two corners of the board 2½" from each side.
5. Hold the shutter firmly against the house and drill into the house through those two holes.
6. Install the corner anchors and hang the shutter on them.
7. Every twelve inches or so around the shutter—2½" from the edge—drill holes through the shutter and into the house.
8. Install anchors in the new holes.
9. Label them on the top and front, and name the shutters.

On wood houses you have to modify this process because the anchor *must* go into the studs and not the siding. I don't know any other way except with some careful measuring to get the holes for the anchors in the right locations.

Impact-Glass Windows

The "glass" part of an *impact window* is strong enough to stop flying debris. There are two types:

- The most common is a laminate of regular glass and a polycarbonate, a very strong plastic (like bulletproof glass). This glass-plastic sandwich looks just like regular window glass. When debris hits it, the glass may break, but the plastic layer keeps the window in place.
- Windows are also available with "glass" made only of the polycarbonate without the sandwich around it. It's normally sold under the brand name Lexan. These windows are also very strong, but you have to be a little careful with them. They are subject to scratching, and some cleaning solutions may discolor them.

(PRO) Impact windows don't make your house look any different than it would with regular windows.

(PRO) Installed properly, Miami-Dade County approved impact windows pass the same tests as strong shutters.

(PRO) The protection is always there. There is nothing to do when a storm threatens.

(PRO) The windows can be great for security, but you often have to improve the locking system for maximum protection.

(CON) Impact windows are more expensive than regular windows plus storm panels.

(CON) And this is big. After a significant hurricane, although the windows will likely have done their job and kept the storm outside, they will be bashed and battered from the flying debris. Some of the glass parts of the laminates will be smashed, so you'll be looking out of broken windows. The frames will be nicked and bent. The windows will have to be replaced—a nasty and expensive construction job. And the likelihood that windows and installers will be readily available is nil. You may be living with broken windows for months or years.

(OPTION) Some people are now installing impact windows but still using shutters. It's double security for intense hurricanes.

(COST) Approximately the same as regular windows plus accordions.

Window Film

Window film is *not* a substitute for shutters. But it has a role to play. Some windows just can't be shuttered or otherwise protected. Other windows—on the upper floors in high-rises facing the ocean, for example—might only need to be strengthened, not protected from debris.

Beware of Bogus Film

All window film is *not* created equal, or anything close. Only the special, extrastrong films are appropriate for hurricane protection. Deal only with reputable companies so you know what you're getting. Can you tell the difference between a roll of hurricane film and a roll of cheap window tinting? I can't. You can try tearing a sample of the film that's being installed to gauge of how strong it is. Hurricane film is almost impossible to tear.

Remember, window film does *not* strengthen the frame or the connections between the frame and the house. Some windows fail because the entire system, frame and all, comes loose. As a result, I don't recommend film for an old, weaker window without doing something to strengthen the frame.

(PRO)   Window film is there all the time. You don't have to "put it up."

(PRO)   You can get tinting in the film to stop the sun from heating up your house and fading your furniture.

(PRO)   A filmed window is much harder to break and provides a measure of security.

(CON)   You have to be careful with filmed windows. The film scratches more easily than the glass.

(CON)   A filmed window is only as strong as the window and frame. Film does nothing to strengthen the frame or its connection to the house.

(CON)   Window film does not meet the Miami-Dade standards, although in Florida you can get a partial insurance discount.

(OPTION) Some film is bound to the window frame by a very strong adhesive. This is a valuable option and can significantly strengthen the window-and-film system.

(OPTION) Film comes in a variety of tints. Don't overtint, however. Your home will be uncomfortably dark on cloudy days and at night.

(COST)   About $10 a square foot installed.

Strengthening and Protecting Doors

Doors may seem strong, but unreinforced garage doors, inward-swinging entry doors, and old-style sliding doors are really very weak and must be strengthened. Because of the size of the opening, a door failure can put your entire house at risk. The sudden influx of air puts a burst of upward pressure on the roof.

There are significant differences in the systems used to strengthen each door opening, so we'll deal with them independently:

Entry Doors, below
Garage Doors, page 241
Sliding Glass Doors, page 242

Permits Protect You

Normally, you can strengthen your existing door without a permit. But if you are contemplating new doors for any part of your house, my recommendation is that you do it right. Get a door with a Miami-Dade County Notice of Acceptance. Then you know it will be strong—as strong as a shutter. And get a permit so you know the door is installed properly.

Entry Doors

The most vulnerable entry doors are:

- Inward swinging doors, especially if they are double doors
- French doors, because the glass makes them weaker
- Doors that are not solid wood or reinforced with metal

All of these door types need to be reinforced to make them stronger against wind pressure, and they need to be shuttered to protect them from flying debris.

Safety Exits

As you plan your door-protection strategy, be sure to give yourself two ways to make a quick exit from your shuttered house. Either choose your two strongest doors to reinforce, but not shutter, or consider the fabric screening system. The screening gives you hurricane protection, but it's mounted away from the house so there is room to get out the door.

If you don't have two strong doors in your house, even with reinforcements, consider replacing your two most vulnerable doors with ones certified by Miami-Dade County. Approved doors are strong enough to stop debris and have robust locking systems.

Reinforcing Doors

Inward-swinging doors are weak because all of the force of the wind gets focused on the one or two latches or locks near the middle of the door. Under high wind pressure, the thin wood holding the locks can easily fail. The idea is to distribute that pressure around the door with slide bolts. Outward-swinging doors are also vulnerable as well because of the suction forces in a hurricane, especially if they are located near the corners of the house. All doors with locks located near the center of one side the door need to be reinforced.

Reinforcing with slide bolts. At a minimum, put a slide bolt at the top and bottom of the door. At a maximum, put another bolt or two on the side of the door that swings in. On double doors, be sure there are bolts at the top and bottom of each side, and then add two bolts between the doors to help connect them together. Be sure the slide bolts go through the header at the top of the door and through the threshold and into the subfloor at the bottom. Don't count on the skimpy slide bolts that are built into the door. Add strong, externally mounted ones.

Shuttering Doors

The most common shutter types for entry doors are storm panels and the flexible screening systems. Some doors can be shuttered with roll-downs as well. Shutter your weakest doors (listed above) first. See the section on shuttering windows on page 221 for a rundown on the different shutter types.

If you have doors that swing in, you can use the type of storm panels that are made to be secured from the inside. The sills that hold the bottom of the storm panels will be removable using "sidewalk bolts" and anchors installed in the cement outside the doors. Remember to give yourself two exits!

Garage Doors

If a garage door fails, there's a big, gaping hole in your house for the wind to enter. Garage doors are now being manufactured that meet the Miami-Dade County hurricane standards. They are somewhat more expensive, and much heavier. But, estimates are that up to 80 percent of the damage to residential homes starts with a garage door failure, so garage-door protection should be a high priority.

If you plan to keep your existing garage doors, they need to be reinforced to better resist the wind pressure and also to protect them from flying debris. Double-wide garage doors are especially subject to failure in strong winds.

Reinforcing Garage Doors

You need to strengthen both the track and the door itself. As a practical matter, thin metal or fiberglass doors can't be strengthened to what, in my opinion, is an acceptable hurricane-protection standard. But all older-style doors can be reinforced to better resist the wind pressure.

An easy test of the strength of your garage-door system is to grab the horizontal part of the track. If you can twist it, your door may pop out of the track and blow in during hurricane-force winds. Reinforcing *retrofit kits* are sold at home-improvement stores. Some garage-door installation companies will do the installation for you.

The weaknesses in the average garage door can be attacked in the following ways:

- Horizontal reinforcing braces are added on the inside of the door to reduce overall flexibility and to keep it in the track.
- Temporary vertical braces are connected to the header and the floor to keep the door from folding or being sucked out by the wind.

- The rollers are changed for stronger ones with longer stems.
- The track is stabilized by connecting it to the wall or ceiling with extra braces.

When extra weight is added to the door, the springs will have to be recalibrated. *This is a job for professionals*. The big springs are dangerous. Don't try to do this job yourself.

It will cost you around $500 to $600 to retrofit a double garage door. Should you put that money toward a new, hurricane-ready door? That's what I would do, especially since you will need shutters for the old doors along with the reinforcements.

Shuttering Garage Doors

If your old garage doors are reinforced as described above, you have gone a long way toward securing that opening into your house. You still have vulnerability, however, because the old-style garage doors are made of a relatively weak material.

The only practical shutter systems for garage doors are storm panels and flexible screening. Either will work well. Storm panels are harder to install because the panels have to be big to cover the door opening. Even aluminum panels are unwieldy in that size. The sill for the panels attaches to permanent bolts in your driveway to secure the bottom of the shutters.

Installing the screening means getting on a ladder to attach the top, but two people can do it fairly easily. The screening also has the advantage that it reduces the wind pressure on the door.

Sliding Glass Doors

Current models of sliding glass doors—especially those with a Miami-Dade County Notice of Acceptance—are significantly stronger than old-style doors. The major improvements are:

- A large lip at the bottom helps keep the door in its groove and prevents water from blowing under the door

- Much stronger frames to keep the door rigid
- Thicker tempered glass or impact-resistant glass
- A larger overlap between the frame and the glass
- Stronger mounting systems

These new doors also seal better to keep out water and noise. And, of course, they are more expensive. Any renovation should include replacing old, vulnerable doors with the new version. In the meantime, old doors' resistance to wind pressure and airborne debris can be improved, although they still won't be up to modern standards.

Reinforcing Sliding Glass Doors

Old-style sliding doors are difficult to strengthen because they are subject to flexing and popping out of the track. Here's an idea, however, that can help:

1. Measure the distance between the sliding section of the door and the inside edge of the door frame.
2. Cut a two-by-four about 2 inches longer than that distance.
3. Put the two-by-four in the wall end of the track and tap it lightly with a hammer next to the slider to force the door closed. You want a tight fit that won't move, but don't overdo it. Trim the wood a little bit if the length is too long. The two-by-four should be at an angle wedged against the door, not lying in the bottom of the track.
4. Do the same thing on the other side of the door if you have a double slider.

Also, a contractor can put steel around sliding doors to strengthen the system. In my opinion, however, old-style sliding doors without the big lips at the top and bottom are dangerous in the hurricane zone. *Never stay in a room with sliding doors during a hurricane.*

Shuttering Sliding Glass Doors

You can use storm panels, accordions, roll-downs, or the flexible screening system to protect sliding glass doors. If the doors are on a balcony, you'll need to get shutter systems that can be closed from the inside, but that is not a problem.

The flexible screen system mounts away from the door, so you have room to maneuver. This system is especially good for older sliders because it lowers the wind pressure that reaches the doors.

Plywood can also be used to shutter a sliding door, although, as always, plywood is not your best choice. Here's how you would do it:

- Build out a wood frame of two-by-fours that fits around the doors.
- Put two or three vertical two-by-fours in the span. Space them about 3 feet apart.
- Anchor the completed frame to the house. The best way is with metal L-brackets and shutter anchors. See page 226.
- Nail the plywood to the frame.

Keeping the Roof On

Since Hurricane Andrew ripped tens of thousands of roofs off of houses, we know a lot more about the important factors in roof strength.

- It's vital to keep the wind outside in order to keep the inside of the house from getting overpressurized.
- The shape of the roof counts for a lot. Moderately pitched *hip roofs*, where there is no gable, are the best style for the hurricane zone. Flat roofs are the worst in terms of aerodynamics. Gabled roofs are also poor aerodynamically.
- The highest stresses on the roof occur near the edges and especially the corners. Extra nails are needed in these areas.
- Roof sheathing should be *nailed* onto the rafters in the hurricane zone. Staples too often miss the wood below.
- Ring-shank nails should be used on all roofs. They hold the sheathing down significantly better than regular nails, for only pennies more.
- Only plywood should be used for roof sheathing. Oriented strand board (a strong kind of particle board) doesn't perform as well.

- Only Miami-Dade County–approved asphalt shingles should be used, and the required installation techniques should be carefully followed. The heavy-duty shingles will last considerably longer, so there is a double benefit.
- Tile roofs provide the best protection from flying debris, and their weight helps keep the roof on. The tiles need to be secured well, however. Miami-Dade County modified the rules for holding tiles to the roof after Hurricane Wilma. Now clips are required on the first row of tiles, and there is better system for securing the ridge tiles. Check to be sure your contractor is using the latest approved system.

Inside the Roof

All houses in the hurricane zone should have:

Hurricane straps. They connect the tie beam(s) to each roof truss. These inexpensive straps are critical. They keep the roof from lifting off. You may have heard of "hurricane clips." Straps are much stronger and insignificantly different in price. Straps can often be added to wood homes after the house is built. It's more difficult to add straps to cement-block houses with concrete tie beams.

Internal roof bracing. An attic that is wide open means the roof structure is not well braced in case of a sheathing failure. The trusses should have their own internal bracing structure, not simply rely on the roof sheathing to keep them vertical. See "Strengthening the Gables," page 246.

Other Roof-Improvement Techniques and Products

Ridge vents. A vent along the ridge of the roof to ventilate the attic space is preferable to turbines that require multiple holes in the sheathing. These vents will keep the rain out as well.

Self-adhesive roofing membrane. This extra layer of protection under the shingles or tiles will prevent water from entering the house if you lose the top layer of roofing. Since this layer doesn't breathe, however, other steps have to be taken to avoid moisture buildup.

Improved soffit design. If the fascia board extends below the underside of the soffit, water is less likely to be blown up into the eaves.

Skylights. Skylights were a problem in Hurricane Andrew, but Miami-Dade County has now approved many new designs. Circular or tubular skylights are a good choice because of their favorable aerodynamics.

Concrete roof. In the hurricane zone outside of the United States it is very common for roofs to be made out of concrete. Concrete roofs have some advantages in warm climates besides their strength to withstand hurricanes. They keep the house cooler, and are not subject to termite damage.

Gluing the roof. In some areas there are companies that will spray a foam glue on the inside of your roof to connect the rafters to the sheathing. This strengthens the roof system by connecting it together.

Net protection. See page 232 for information on a new product that can help strengthen weak roof systems. A net is draped over the roof and anchored to the ground.

Strengthening the Gables

In some areas concrete-block houses are built with wooden gables. This can be a significant weak spot. If the gable isn't reinforced, it can blow in like an unlatched door. The only way to know where you stand is to go into the attic. If the gable is made out of wood, you should see angular braces connecting the gable to the nearby roof trusses. If there is no significant bracing, call an engineer or architect to determine how to fix the weak-gable vulnerability.

Gable vents. Large, weak gable vents can also be a weak spot, even if the gable itself is strong. Consider replace flimsy vents with stronger models or plan to cover the vent with plywood or a shutter during a hurricane.

Making a Stronger House

Tie It All Together

It is critical that houses in the hurricane zone be tied together, from the roof down to the foundation. In Hurricane Andrew, even some concrete block houses collapsed because the walls lacked vertical columns. The idea is to connect the roof to the tie beams with hurricane straps; the tie beams connect to vertical cement-filled columns containing steel rods; the columns connect to the foundation. The house is one unit. In properly built wood houses, a similar concept is used.

The vertical steel bars in the poured concrete columns spaced around the perimeter of the house add dramatically to the strength. Miami-Dade County requires poured concrete around the windows as well.

Look for metal hurricane straps in the attic looping over every truss as a first step in determining how hurricane-ready your house might be. You'll need an inspector, however, to get the full picture.

Keep the Water Out

A waterproof elastomeric coating can be applied to concrete-block walls to seal the water out. Even if cracks develop, this coating will expand to seal the opening.

Also, a properly designed connection between the walls and the floor slab in a concrete block home will direct water away from the house in a heavy, wind-driven rain.

Cast Concrete Walls

Walls made of solid concrete are significantly stronger than concrete block walls.

Concrete Second Floors

For years homes have been built with concrete-block first floors and wood second floors. The second floor ends up being significantly weaker. This style of construction makes no sense, since the second floor is subject to stronger wind. Concrete blocks should be used all the way to the roof.

Backup Power Solutions

For most people affected by a hurricane, the most aggravating problem is the loss of power. In the hurricanes of recent years, millions of people suffered through days and weeks without power from the electric utility even though they didn't have significant damage to their property.

Having power after the storm requires careful planning well in advance. Here are some of your options:

Generators

There are two kinds: portable generators and permanent standby generators. *Portable generators* can be carried or wheeled around and normally run on gasoline. *Permanent standby generators* are permanently attached to your house and generally run on natural gas or propane.

Depending on the size of the generator you choose, you can run your whole house or just a selection of your household appliances. Small generators will let you run some lamps, a TV, your cell phone, and laptop, and perhaps a small air conditioner and refrigerator, while big permanent units will run everything, including the central air. See page 255, "Generators," for the pros and cons of each type to help you choose.

In general, generators are *not* an option if you live in an apartment or condo.

Battery Backup Power Systems

How I Do It

I live in a condo now, so my backup system of choice is portable battery power. I take two of them to work with me, where we have a big generator, and charge them during the day. They work all night, with plenty of power to spare, charging my cell phone, running the laptop, and powering some fluorescent lights, a TV, and a fan.

For between about $80 and $300 you can buy a portable battery power system. They are nothing more than a sealed battery inside a plastic case with an attachment called an inverter to convert the DC battery power to AC for your appliances.

They have two or three outlets on the side, and the ones you can carry usually have a light built in. In general, the more expensive they are, the more electricity they hold.

In my experience, they charge pretty quickly and last a long time if you use common sense about what you try to power.

I prefer to have three or four of them, instead of one big one. That way, each member of the family can have his/her own.

Recharging battery systems. Battery systems, obviously, need to be recharged. You'll probably want to do it daily. If you can find convenient AC power—your neighbor with a generator or your workplace, for example—that's one way to do it.

Your other option is to find solar panels that plug into the charging socket. I've seen them in the past, but have had trouble tracking them down recently. If you have any information on an inexpensive portable battery system that easily recharges with solar panels, please let me know at www.hurricanealmanac.com.

Heavy-Duty Extension Cords

It often happens that your neighbor gets power back, but yours is out for some time. The service to the neighborhood may have been restored, but a

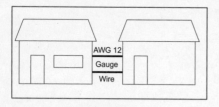

damaged line between your house and the pole is keeping you in the dark. If you do it right, you can safely use some of your neighbor's power.

You *must* use large-gauge extension cords only. NEVER STRING HOUSEHOLD EXTENSION CORDS TOGETHER! I recommend on the shopping list that you invest in at least two outdoor fifty-foot AWG 12 (American Wire Gauge 12) brightly colored cords.

Do It Safely

- Add up the power used by everything plugged into each cord. If you're adding amps, the total should not be over about 7 amps, or about 800 watts if you add up the watts.
- Routinely have your neighbor check the plugs where your cords go into his receptacles to be sure they are not getting too hot. If they are, don't power as many things at the same time.
- If you string more than one cord together, tie the cords in a knot where you plug them into each other. If someone stumbles over the cords, they won't come unplugged.
- Use common sense. Keep any connections out of the water.

If you can, run more than one cord from your neighbor's house. Connect each one to a different circuit. Don't run the extension cords for more than 150 feet total. With cords from two circuits, you can live somewhat comfortably, even have some air-conditioning. See "Can I Have Air-Conditioning?" page 264.

MY DISCLAIMER: *Always follow the precautions listed on the extension cord package.*

Power-Saving Lights and Appliances

You will "enjoy" the no-power or low-power period after the hurricane a lot more if you have some energy-efficient appliances around the house or apartment. The difference in the amount of time that you can use your backup battery system or your portable generator without recharging or refilling will be *dramatic*. Unless you have a big generator, investing in some energy-saving devices could make your life a lot easier.

Lighting

Incandescent. Use *no* incandescent or halogen lights, that is, any lights that get hot. You are wasting a lot of energy generating all that heat. Your two choices for efficient lighting are fluorescent and LED.

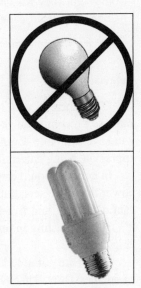

Fluorescent. Replace your screw-in incandescent bulbs with comparable fluorescent bulbs and you'll save 65–80 percent of your lighting electricity. That's a *lot*.

You can find a compact fluorescent light bulb (CFL) that fits in just about any lamp. You'll spend more for it than for an incandescent bulb, but it will last for years and uses so much less electricity that you come out ahead in the end.

LED lights. Bulbs made of LEDs (light-emitting diodes) are relatively new and even more efficient than fluorescents. These will save you about 90 percent of your electricity, plus or minus. In most cases, they aren't as bright as fluorescents and tend to have directed light, as opposed to fluorescent and incandescent bulbs that throw light in all directions. That means LEDs are good as reading lights and for other applications where you need to light a confined area.

Lighting at home. My suggestion is that you get a few LED lights for reading and otherwise light your home with fluorescents.

Flashlights. Consider replacing the lightbulbs in your handheld C- or D-cell flashlights with LED bulbs. They will last ten times as long, over a week in normal use. They are not cheap—$8–$10 a bulb—but they will give you peace of mind and they last "forever." Check www.HurricaneAlmanac.com for more information.

Television

Often there is very little to do during the dark nights after a hurricane, so I recommend you get a TV that will work with whatever power you have available. Even people who don't watch television very much appreciate being entertained and knowing what's going on.

In general, liquid crystal dislay televisions use half to one-third less power than tube sets, but there is a lot of variability. Still, that's significant over time, and five-inch LCD sets use very little power, so I think it's worth investing in one of those regardless of what you do for your "big" set.

If you are going to have limited power, think about getting a 15- or 17-inch LCD set. That's big enough for the family sit around and watch, but small enough to run for a long time on a small generator or batteries.

(DO IT NOW!) Before you forget, connect rabbit ears to your new LCD set. Experiment to see what you can pick up off the air, so you'll know what you're going to have to do when the cable goes out.

Television sets with picture tubes use more energy than models with LCDs.

Refrigerator or Thermoelectric Cooler

Most refrigerators use lots of energy, but you have options, depending on how much power you have and how much "cold" you need.

A small refrigerator will use around 200 watts, plus or minus, so unless you get an air conditioner, the fridge will be your biggest single energy consumer. Even a relatively small generator will keep it going, and you'll be able to make your own ice.

If you can live with cool beverages and food and forego the ice, there are thermoelectric coolers or travel coolers available. They cost from less than $40 to $120 (depending on the size), use relatively little electricity, and will keep food warm or cool, depending on how you set it. Coleman makes a unit that looks just like a small refrigerator. Google "travel cooler" and you'll find a lot of them.

Computer

Laptop computers use considerably less power than a desktop plus a monitor, so that's the way to go if you can. The exact difference has to do with the size of the laptop and the type and size of the monitor, among other factors.

If you're using a generator, especially a small one, you'll want to consider an uninterruptible power supply (UPS) to isolate and protect the computer. Some generators produce what's called "dirty power," which can harm a computer. (See "Generators," page 255.) If you're using a battery-powered system, a UPS can still help you by alerting you when the electricity in the batteries is running out.

NOTE: *You cannot use just any UPS with a generator!*

Many UPS units will see the distorted power from the generator and, for safety's sake, stop accepting it. You must determine whether your UPS is "generator compatible." Look carefully at the specifications. Some UPSs will allow you to adjust the sensitivity of the unit so that it allows dirty power to flow in, keeping the UPS battery charged.

Generators

A generator can make life bearable after a hurricane. Unless you are very lucky—as we were in southeastern Florida after Wilma—the weather is going to be hot and humid after the storm passes. Picking the right generator, however, can be complicated. And not using it right can be deadly.

Warning!
A dozen people or more have died in Florida in the past few hurricane seasons from carbon monoxide poisoning! But with care and common sense, owning a generator can be a good experience.

Portable generators come in many sizes and shapes.

Permanent standby generators look like outside air-conditioning units.

There are two types of generators: *portable generators* and *permanent standby generators*.

Portable generators are more dangerous because inexperienced people do inappropriate things with them. But still, most people will opt for a portable generator because they are much cheaper.

Portable Generators

Here are the rules for using a portable generator, never to be broken:

- NEVER run a generator *inside* the house, garage, or an enclosed porch.
- NEVER run a generator *under* an overhang or in an open porch.
- NEVER connect a portable generator to your house's power system.
- NEVER refuel the tank when the generator is running. Leave it off for fifteen minutes before fueling.
- ALWAYS use heavy-duty outdoor extension cords.
- ALWAYS buy a couple of battery-powered carbon monoxide detectors at the time you buy the generator.
- ALWAYS keep the generator dry (which often means you have to turn it off and protect it from the rain).

The Two Worst Mistakes That People Make
- They put the generator too close to the house, where the colorless, odorless carbon monoxide gas can seep inside.
- They try to power the house circuit and unintentionally electrocute a power worker.

Please, follow the rules given above.

Choosing a Portable Generator

Portable generators are, in a word, portable. You store them in the garage and move them outside when the power goes out. Then you run heavy-duty extension cords into the house to bring electricity to what you want to plug in.

The biggest downside to a portable generator, versus one that's installed in your house, is the gasoline that portables require to keep going. Nobody wants to have a lot of gasoline near their house during the hurricane, and getting gasoline after the storm can be a problem. It presents a dilemma when you are choosing what size generator to buy.

The generator dilemma. Bigger generators power more stuff but, in general, use more gas. "In general," because some generators are more efficient than others. The difference in efficiency can be dramatic with a cost difference that is not very much, so look closely at the specifications before you buy. The important calculation when comparing efficiency is: *Divide the number of gallons in the tank by the time the generator will run at 50 percent load.*

For example, Coleman says their 5000W Premium Plus unit will run 11 hours on 5.6 gallons of gas at 50 percent load, or about 0.5 gallons per hour. Honda says their Deluxe 3300W unit will run 15.6 hours on 6.6 gallons. That comes out to 0.42 gallons per hour. Smaller capacity generators are usually more efficient.

Electricity Versus Gasoline Consumption

The question is, how much electricity do you need versus how much gasoline is it practical for you to store or get? But before you can add up your total wattage, there's another wrinkle to consider. Notice that the generators I mentioned were rated at 50 percent load. That's because, while a 5,000-watt generator can indeed produce that much output power, it is significantly less efficient at full load. Also, appliances like air conditioners and refrigerators and even TV sets with picture tubes drain extra power when they start up. So you have to account for that.

What size portable generator? My suggestion, to give you a good margin of safety, is that you add up the consumption of all of the things you want to power (ignore the start-up power), then multiply by 2. That will tell you how big a generator to get. The start-up surge from the air conditioner or refrigerator will be taken care of by the extra capacity you have, and under normal operations you'll be running at maximum efficiency.

Let's look at two energy budgets using standard items on the left and energy-saving items on the right. (See, "Power-Saving Lights and Appliances," page 251.)

Two Energy Budgets Compared (in Watts)

	Standard	Power-Saving
Five lights	500 (incandescent)	125 (fluorescent)
TV	200 (27")	50 (17" LCD)
Laptop	100	100
Medium-size refrigerator	500	
Travel coolers (no ice, but cool drinks and food)		100 (50 watts each)
Air conditioner	1,100 (12,000 BTU)	800 (8,000 BTU)
Other	100	75
TOTAL	2,500	1,250
GENERATOR NEEDED	5,000 watt	2,500 watt

The wattage figures are estimates, but I hope you get the point. If you work hard to limit your power consumption, you can spend less money on a generator and use less gas. The biggest benefit of being a power miser may turn out to be limiting the aggravation of getting gasoline.

If you are able to use a smaller, window air conditioner and are very careful with your power use, you can even go to the next smaller size generator, around 1,500 watts, and be comfortable.

Some large portable generators have a 220V outlet. That means (if you are very careful about the total load you're plugging in) you can temporarily power your clothes dryer, which may be a godsend in the aftermath of a hurricane. Since the generator has to be outdoors, you'll need a special heavy-duty 220V extension cord to run to the dryer.

Storing Gasoline

It's not safe to keep a lot of gasoline around your home, and it should *never* be kept inside the house. Inevitably, however, you're going to have to store some gasoline, so here are the rules:

- *Only* use Underwriters Laboratories–approved *red* containers. They come in 1-, 2-, and 5-gallon sizes.
- Only fill the containers 95 percent full to allow the gas to expand when it gets warm.
- Put the caps on tight.
- Store the containers out of the direct sunlight.

Try to get your gas supply right before the storm. Gasoline will deteriorate in temperatures over 80°F, which, of course, they usually are during hurricane season. If you have gasoline left over after the storm, use it in your car and get new gas when another hurricane approaches.

Storing Your Generator

To be sure your generator will work when you need it the next time, get some fuel stabilizer from the auto supply store or gas station.

Fill the generator tank about halfway and pour in the appropriate amount of stabilizer. (Read the label.) Start the generator and let it run five to ten minutes to be sure the stabilized fuel gets into the engine. The stabilizer keeps the gasoline from getting gummy in the warm weather. As a precaution, you might want to run the generator a couple of times before the next year. Because your generator is full of "summer gas"— refineries mix gasoline differently for different seasons—it will run best on a warm day.

Permanent Standby Generators

These systems power all or part of your house, and usually kick in automatically when the power goes out. Normally, they look a lot like the part of a central air conditioner that sits outside. Most home units use either natural gas (delivered from a gas company as a utility) or propane (LP) gas with a tank or tanks that are periodically filled.

Even the cheapest permanent standby generator installation—powering only part of the house—is thousands of dollars more expensive than a portable generator, even if you add on the cost of a portable air conditioner, LCD TV, and other energy-efficient items. Still, if you can afford it, it's a wonderful luxury to have full power on a continuous basis.

You must have a professional involved in the purchase of a permanent standby generator. You're going to need a licensed electrician for the wiring, installation, and required inspections to be sure it's done right. Be safe. Don't scrimp.

To choose the right permanent standby generator, the first question is "Do I need to power my whole house, or can I cherry-pick what I want to use after a storm?" Some people just want to power the air conditioner, refrigerator, and smaller appliances, and will do without the stove, dishwasher, and the rest. They'll cook on the grill or on a camping stove.

Others want to be able to use everything in the house, including the washer and dryer. That requires a much bigger generator, which in turn uses more fuel. Fuel can be an issue if the gas comes from propane tanks. Eventually they'll run dry. If you have a natural gas connection from the utility, the main issue is money. It's expensive business powering a whole house on natural gas.

Regardless of whether you decide to power your whole house or part of it, here are the rules:

- Use only a licensed contractor.
- Check with your local building department to see if you need a permit. An inspection will assure the installation is done correctly.
- Install a *transfer switch* that either automatically or manually switches between your power company and the generator. This is an important safety measure. Note: The permit process can take a month or more, so make your application early.
- When the generator is installed and working, notify your power company that you have one. That way they will know that lines in your house will be electrified even though your neighborhood is out.

Generators and Home Electronics

Inexpensive generators, in general, make what's called "dirty" power. Technically, it means that the shape of the wave of the AC current is not smooth. The offshoot for you is that some sensitive home electronics may not work well on generator power.

Computers, especially, should be isolated and protected from the generator. The best way to do that is with a UPS (uninterruptible power supply). But BEWARE! Not every UPS will work with dirty power. See page 254 for more information.

Computer Hurricane Plan

If your whole life is on your computer, you need to have a computer hurricane plan to go along with a plan for everything else. Here are some tips to keep your computer going and to protect your pictures and other files.

Back Up and Back Up Again

You need at least two backups of your important files. It's easier than you think. Here are some of the options. You can choose any two that work for you, depending on how many files you have.

- Get an *external hard drive* that plugs into the USB port on your computer. For most people, an 80 GB drive is big enough, but they come much larger if you need more capacity. These days you can get a 250 GB external drive for less than $150. That should be enough for anybody. Software will come with the drive to help you back up your files, or you can simply make a copy of the folders holding your pictures and other important files. If you've set up your Windows software right, it should be as easy as copying the "My Documents" folder. But don't forget about financial files and other things that might be outside that folder.

- If your computer has a *CD or DVD burner*, back up critical files that way.

- You can buy a *USB Thumb Drive* that will hold from 2 GB (for $50 plus or minus) to 8 GB (for around $160). It's a quick and easy backup alternative if you don't have too many files, although not the cheapest. Other systems listed here will get you more capacity for less money.

- *E-mail backup*. E-mail photos of your house to somebody out of town to be sure you have them for the insurance company. You can also use a free e-mail account as a backup location for all kinds of files. You can now get up to 2 GB or more of free space. That will hold a lot of data. Yahoo!, Hotmail, Google's Gmail, and AOL all offer free storage. Once you have one or more accounts, just e-mail your important stuff to yourself. Some free accounts expire if you don't use them, so they may not be good for permanent storage. Still, they'll get you through the storm.

Put your backup in the lockbox with your most important papers. If you can, send a backup CD, DVD, or Thumb Drive to a relative out of town.

Internet

DSL versus cable. In my experience, DSL is normally more reliable than cable Internet in a storm. Phone wires (which carry DSL) are more robust than the cable TV distribution system. Still, you should have a dial-up account as a backup.

NOTE: If you get your DSL or cable Internet through Earthlink, for instance, you may get twenty hours per month of dial-up for free. You can use this for your Internet connection if you have to evacuate. Other companies may have a similar deal.

Wireless Internet. An alternative system that *should* keep you connected longer is the wireless system for your laptop now offered by providers like Cingular, Sprint, and Verizon. For $50 to $60 a month, if you have cell phone service with the same company, you can get reasonably high-speed service in much of the country. Not every company serves every area, however, so you'll have to ask which, if any, is best for you. Also, some cell phones can be connected to your computer to get an Internet connection.

Power

Getting good information in the hours leading up to a storm can be difficult. The Internet may turn out to be your best source, depending on where you live. So, keeping the Internet going could be a high priority.

You'll be able to use the Internet much longer if you have one or more UPSs for your computer equipment. These units—which have batteries in them—are also your best protection against power surges. My suggestion is that you have two: one to power your DSL or cable modem and router (if you have one), and the other to power your computer. That way, if one fails for any reason, you have another working unit, and the modems and router will work longer.

A laptop is the best because it uses less power and won't use up the UPS battery as quickly as a desktop. Also, it has its own battery, obviously, so you can be online for another couple of hours or so after the power goes out.

Plug as little equipment into the "battery backup" sockets on the UPS as possible; that way your power will last longer. Your printer, speakers, and other accessories only need surge protection. During a hurricane, you're not likely to need them anyway.

An annoying fact of UPS life. Most of them will beep incessantly when the power goes out, like the fact that the house is dark isn't notice enough. There's apparently not any way to shut that off on my UPSs. You might want to consider having a heavy-duty extension cord ready so you can put the UPS in a closet or another room.

NOTE: You could run into a problem using a UPS with a portable generator. Be sure your UPS is generator compatible before you buy it.

Move Your Computer

Don't take a chance on your computer getting wet. Be sure it is nowhere near a window when the wind starts blowing, even if you have shutters.

Your Ideas

If you have any other tips for keeping data and computers safe, please let me know at www.hurricanealmanac.com.

Can I Have Air-Conditioning?

The answer, of course, is yes, if you are willing to pay to have enough electricity to power it. But you don't need a whole-house generator to keep at least part of your house cool.

If you're going to buy a portable, gasoline-powered generator, you are not going to be able to run your whole-house A/C. The options, depending on the type of windows you have in the house, are to have a *window unit* or a *portable unit.*

If you have *single* or *double-hung windows,* that is, windows that slide up and down so they can easily accept a window air conditioner, you can use either type. If, however, you have awning windows or other types that won't take a window unit, you'll only be able to use a portable unit.

How I Did It

When I lived in a house in Miami, I installed a 6,000 BTU air conditioner in my master-bedroom wall. I positioned it behind some curtains. When the power went out, I got out my portable 1,500W generator, strung a heavy-duty cord outside, and ran the unit until the room was cool. Then I then shut off the air conditioner and ran a fan on my portable battery power system while I slept (I charged the batteries with the generator at the same time I was cooling the room). I didn't like the generator noise bothering me and the neighbors all night. It worked perfectly. I'd restart the air conditioner in the morning and run it until I left for work.

A *portable air conditioner* has wheels and a tube coming out the back for expelling the hot air outside. The tube has to go out through a window or door, and that requires some advance planning, depending on the type of windows you have in your house.

How Many Watts?

To calculate the power consumption of an air conditioner, take the number of BTUs and divide by the EER (energy efficiency rating). For example, if it's a 10,000 BTU unit with an EER of 11.0:

10,000 BTU ÷ 11.0 EER = 909 watts

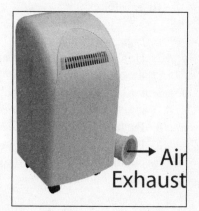

Air Exhaust

The obvious advantage of a portable unit is that you can take it from room to room, depending on the time of day. A 10,000 BTU unit sells for less than $500, and will cool a good-sized room.

Be sure you get extra-heavy-duty extension cords to run from the air conditioner to the generator. Also, check the specs to be sure your generator has enough capacity to handle the power surge when the compressor starts up.

Candles—The Right Way

The "official" word from the fire department is "Use ONLY flashlights before, during, and after a hurricane because we may not be able to respond." Indeed, there are examples of candle accidents during storm events that resulted in the total loss of a home.

But we've got to be realistic. Some people are going to run out of batteries and will then need to use candles. It's going to happen, so it makes sense to know how to do it with maximum safety.

First, use candles *only after* the storm has passed and the wind has completely died down.

Second, use VOTIVE CANDLES ONLY, because they are short and squat. And *always put them in a glass container in the middle of a large, nonflammable dinner plate*.

To be safe:

1. Light the candle.
2. Drip some wax into the middle of the plate.
3. Push the votive candle holder into the wax to hold it firm.
4. Put the candle in the holder.
5. Put the plate and candle in the middle of a table, chest, or counter where there's no chance it can be knocked off.

Keep candles away from breezes, young children, and pets. In fact, if any are present, you should stick with flashlights.

The Disclaimer

I repeat, the fire department recommends that you NOT use candles, and, obviously, there is some potential for problems. DO IT AT YOUR OWN RISK.

Pool Preparation

Take a few sensible steps before the hurricane to prepare your pool and it will be much easier getting it back in operation after the storm. You'll notice that a few of the steps I've listed are different than those you have seen or read elsewhere. I'll explain at the end of the section.

The Do's and Don'ts

- DO NOT lower the water in your pool.
- DO NOT throw furniture or anything else in the pool.
- DO turn off the circuit breaker that goes to your pool pump.
- DO wrap your pump in plastic and tape it well.
- DO bring *all* of your pool supplies and furniture inside.
- DO superchlorinate (shock) your pool.

Superchlorinating the Pool

Ask your pool expert how to superchlorinate your pool. The instructions vary, depending on the size of the pool and the type of chlorine you use. A rough rule of thumb is to put double the normal chlorine in the pool the evening before the hurricane is due. That chlorination should last a while since it's not likely to be sunny for a day or two.

Dispelling Pool Myths

DO NOT lower the water in your pool. If it rains over about six inches (the normal level of the water below the lip), your pool will fill up and overflow. After that, any rain that falls over the 500 or so square feet of the surface of the pool will run onto the deck, into the yard, wherever it will. This is just like the rain that falls on the driveway, the street, the patio, or the roof. None of them absorb water, either.

If you drain your pool too far, the pressure of the ground water pushing up can cause significant structural damage. If the hurricane doesn't come with a lot of rain, you have to fill the pool back up using the hose, and the municipal water may not be working. This is why I recommend, on balance, that you not drain the pool.

DO NOT throw anything in the pool. Concrete pools are made of a soft material that scratches very easily. Vinyl pools can be punctured. Neither is made to hold sharp-edged objects. Also, since you're super-chlorinating the water, the items will be bleached and may be damaged by the high chlorine levels. I recommend you store your pool furniture and supplies in the house or garage.

Your Pet's Hurricane Plan

Don't forget your pets when you're planning for a hurricane. Many of the issues are the same as those you face for yourself.

The Most Important Thing

Have your pet's vaccinations up-to-date at the beginning of hurricane season. Consider getting a microchip implant for all pets. At the very least, have tags and a good collar for dogs and cats.

Where Will Your Pet Ride Out the Storm?

If you will be evacuating, you need a plan for your pet. The options are:

- *Take the pet with you.* If you're going to a hotel, pick one that takes pets.
- *Board the pet at a kennel.* You'll need to have a reservation and take the animal there early.
- *Go to a shelter that allows pets.* Some communities have pet-friendly shelters. Check the policies in advance. You are normally restricted to dogs and cats, and you must bring them in cages. Also, you'll probably have to register in advance.
- *Leave the animal at home.* Sometimes this may be your only option. Prepare a room or rooms with plastic sheets on the floor, lots of food and water, and as many pet comforts as possible. Don't tranquilize the animal.

NOTE: Never leave a cat and a dog together, even if they are best friends. Bad things can happen under the stress of a hurricane. This same rule applies to other types of animals as well.

If You Stay Home with Your Pets

- Bring pets in the house. Don't leave them outside or in a shed.
- Falling barometric pressure can cause an animal to react in an unexpected way. Some dogs and most cats are best kept in a carrier. Birdcages should be covered.
- Have a pet carrier for every animal in the household.
- Be sure every animal has identification. Tattoos and microchips are the best.
- If you ever have a reason to open the door (normally a bad idea), be careful. Confused pets can dash out unexpectedly.

Pet Hurricane Supplies Checklist

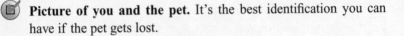

Pet carrier. Get one with plenty of room for your pet to move around. Collapsible models are the easiest to handle.

Bowls, food, and water. Bring your pet's regular bowl, two week's supply of dry food, and at least five gallons of water.

Picture of you and the pet. It's the best identification you can have if the pet gets lost.

Proof of vaccinations. Take the paperwork with you if you evacuate.

Litter pan, litter, and newspapers. Take whatever your pet is going to need with you. Pets need a potty, too.

Aquariums in a Hurricane

If you have an aquarium, especially a large one, you need to make special preparations for the loss of power. There are three things you need to be concerned about:

- Keeping the water oxygenated so the fish can breathe
- Removing the waste the fish produce
- Keeping the tank's water at the right temperature, especially if you have tropical fish

A generator is the obvious answer, but that requires that you have enough fuel to be sure the power is not off for more than a few hours. You can also get battery-operated air pumps and you can rig up a filtering system. Heating the tank is a challenge if that should be necessary.

In any case, the day before the storm consider taking the following actions:

- Change the water and clean the tank.
- Do not feed the fish much, if at all.

- Have your backup plan in place and ready to go if the power goes out.

The bottom line is: Fish need a hurricane plan, too! Check with an expert so you do the right thing for your type of fish.

Your Boat's Hurricane Plan

> ### Your Life or Your Boat?
> In hurricane after hurricane, people die on boats. Don't let that happen to you. Have a plan and stick to it if a hurricane comes. Under no circumstances should you ride out the storm on your boat, no matter how big it is.

If your boat is going to ride out a hurricane successfully, you need to have a good plan. Boats require *early* preparation, well before the storm affects the local weather.

Preseason Planning Checklist

Check your insurance and marina rules. Determine whether you will have to evacuate your boat for a storm.

Know what your options are for harboring the boat. Are there local marinas and waterways that give your boat a better chance of surviving the storm?

Inventory or videotape the contents of your boat. Seal the tape in plastic and keep it with your valuables or put it in a safe-deposit box.

Make copies of any documents that you routinely keep on the boat. Keep them in a safe place.

Under a Hurricane Watch Checklist

In many situations you need to make full preparations under a hurricane watch, even though it's not at all certain the storm will come. Gauge your time by considering what you'll need to do at home and at your business.

 Move trailered boats to high ground. Take half of the air out of the trailer tires. Remove everything possible from the boat. Lash it down to anchors driven or screwed into the ground.

 Move in-water boats early if you have to evacuate. Don't get caught in a crush with other people doing the same thing, or in deteriorating weather.

 Remove everything possible from a boat left at anchor, including radios and accessories. Don't forget to take all paperwork with you.

Tie the boat so it can rise with the water. Use spring lines with half-hitch knots. Put the lines high on the pilings. Use extra lines.

Business Hurricane Plan

Hurricanes are tough on everybody, including businesses. But many businesses suffer unnecessarily because of poor planning. A good business hurricane plan can make a difference in how—or whether—you continue operations after the storm passes.

Failure to Communicate
The biggest hindrance to getting businesses operating after a storm is the inability to communicate with employees. If you don't do anything else, set up a communications plan.

Communications plan. Make a plasticized card for every employee with key phone numbers on it. Include an out-of-town emergency number where employees can check in, let management know about their personal situation, and get instructions.

Employee emergency profile. Have supervisors privately question each employee to learn their personal and family hurricane situation. Everyone should fill out a form answering questions about family size, possible handicapped or elderly dependents, and so on. This will help you determine the size of the workforce you might have after a major storm.

Backup, backup, backup. Any and all electronic records need to be backed up to a location outside of the area. This should be routine procedure anyway, in case of a fire. Scan any critical papers and include them with your backup files.

Phone backup. Contact the telephone company to set up a system for transferring your main phone lines to another location or to voicemail so you can stay in touch with customers.

Hurricane supplies. Stock your business with hurricane supplies that you might need after the storm.

Shutters and hurricane protection. Have a trained inspector look at you place of business so you know its weaknesses. Shutter or protect all windows and doors.

Inventory and equipment protection. Move any merchandise or valuable equipment off the floor. If possible, store anything of value on shelves or on concrete blocks. What you can't move, cover with plastic or tarps.

Clear desks and walls. Put away anything that would blow around if wind got into the building.

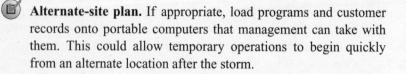 **Alternate-site plan.** If appropriate, load programs and customer records onto portable computers that management can take with them. This could allow temporary operations to begin quickly from an alternate location after the storm.

Management hierarchy. Establish a clear order of succession in management so that decisions can be made, emergency supplies can be bought, and vital issues can be addressed if one or more key personnel are incapacitated or unavailable.

Management sensitivity. Be sensitive to your employees' fears and needs before and after the storm. You will have better, more loyal employees in the long run.

Where to Park Your Car

Cars are often a problem during hurricanes. Unless you have a garage with a strong door, it's hard to find a place for a big thing like a car where it's not going to be affected by debris blowing in the wind. Here are some ideas:

Parking garage. This is your best option if you have one nearby that will stay open during the storm. If you have two cars, consider putting at least one of them in a parking garage above the ground floor. The wind does *not* go through the garage at full force, so your car should be safe.

The City of Miami allows residents to park in city garages for free during hurricanes. All municipalities should do that.

Stay away from trees. Find a place to park your car where a tree can't fall on it, even if you have to walk a few blocks to get home. Thousands of people are now wishing they had thought of that before the hurricanes of the past couple years.

High ground. Every time there's a hurricane we see cars flooded up to their windows. The solution is not complicated: Move your car to high ground. If you have to leave it outside, park it in a place where it's not likely to be flooded and hope for the best.

Behind a building. Your car's best chance of surviving in tact is on the downwind side of your house or a nearby building. If you can figure out the direction the strongest wind is going to blow *from*, park the

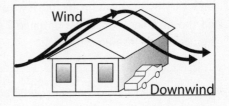

car sideways on the opposite side of the house. Park as close to the center line of the house as possible. If you park close to the house and are fortunate, any debris flying off the roof will likely miss the car during the worst of the storm.

Hurricane Insurance

As a result of the hurricanes of the past two years, hurricane insurance is getting more expensive and harder to find. Private companies are limiting new policies and reducing exposure. On page 176 I explain why the hurricane insurance system is never going to be stable without federal intervention. For now, however, there is no good news.

If you own your home, there are three parts to your insurance coverage:

Flood insurance. The National Flood Insurance Program (NFIP) was introduced by the federal government in 1968. This, essentially, made flood insurance a taxpayer-subsidized program. You buy a separate policy from your insurance agent and pay a separate premium. Coverage of damage from rising water is, therefore, excluded from your homeowner's policy. You can save money on a flood policy if you get a special elevation report on your property. Call 1-888-FLOOD29 for more information.

Homeowner's insurance. In many areas your homeowner's insurance covers damage from everything except flooding caused by rising water. If your toilet backs up, homeowner's insurance will cover you. The deductible is typically about $500. Increasingly, however, wind damage from hurricanes is excluded from these policies. The details may vary by the location and the laws of your state. There are questions on the Insurance Checklist on page 277 to help you understand your homeowner's insurance coverage.

Shutter Discounts

In Florida and some other states, insurance companies are required to give discounts to people who strengthen their property with shutters or other hurricane protection systems. Different companies give different discounts. That's a good reason to shop around, if you have options.

Wind insurance. Separate policies to cover wind damage—with high deductibles and high prices—are cropping up in more and more coastal areas. Deductibles are typically 1 to 5 percent of the value of the structure (not including the land). In Florida, two-thirds of the homeowners now pay a 2 percent hurricane deductible. Therefore, on a $200,000 house, the homeowner pays for the first $4,000 of damage.

The details of wind coverage vary from place to place and policy to policy. Go over the questions below with your agent so you understand your coverage.

Renter's Insurance

There are comparable policies for renters, except you only have to insure your contents. It is the landlord's responsibility to insure the structure itself. The landlord's policy(s) will *not* cover your belongings. Call an insurance agent and protect yourself.

Auto Insurance

Normally, your car insurance will pay for hurricane damage to the vehicle. If your car is in the garage and the roof collapses, there might be a fight between insurance companies, however. So it may be a good idea to have your policies with the same company.

Insurance Checklist

> ### Know Your Coverage—Ask Your Agent
> Insurance is annoyingly (and unnecessarily) complicated. Ask your agent so you know the answers to these questions.

 Am I covered for hurricanes? In many areas, hurricane damage is *not* part of your homeowner's insurance.

 Am I covered for rising water? Homeowners insurance *never* covers flood damage from rising water. You need a separate flood insurance policy.

 Do I have "replacement value" insurance? In some places it is required, but not all. You want to know what happens if your ten-year-old sofa is ruined. Are you going to get a new, comparable sofa? Or are you going to get the value of the old sofa, which is likely not very much? Do you have "inflation insurance" so that you will get a comparable sofa at today's prices?

 Do I have enough insurance? You want to know that your house or condo will be replaced if the worst happens. Property values and construction costs have increased over the past few years. Be sure you're still covered. The insurance that covers the cost of bringing your property up the current code is called "ordinance or law coverage." You need it.

 What is my deductible? In many places, hurricane deductibles can be quite large. Often 2 to 5 percent of the insured value of the property. You want to know how much you would have to pay out of your pocket if something bad happens and you have

$50,000 worth of damage to your home or apartment and its contents.

 Who pays if I can't live at home? Many policies will pay some or all of your living expenses if your house in uninhabitable. You want to know how much and under what circumstances the insurance company will pay so you can determine what your options are.

 Who pays if I have to store my possessions? If your home or apartment is damaged, you may have to put your possessions that survived in storage while you live elsewhere. You want to know who will pay for that.

 Are my computer, artwork, firearms, silverware, and jewelry covered? You many need to pay extra to get coverage for special items.

What else is not covered in my house or apartment? You want to know about any exclusions. For example, who pays if the sewer system backs up into your house after the storm?

Who pays if my car gets damaged in the garage? You want to know who pays if the roof of the garage falls on your car, or something bad happens in the condo's garage.

If You Have a House

What's not covered in my yard? Quite often your fence, shed, and other items outside your house are *not* covered. You want to know where you stand. Maybe you'll secure the uninsured items better before the storm.

If You Have a Condo

What part of my unit is covered by my personal policy? In general, your hurricane policy covers everything inside the rough walls that surround your unit. That includes all of your possessions, the kitchen cabinets and appliances, the light fixtures, and so on. You want to know exactly who is responsible for what. How about the walls, the windows, the doors, the balcony screening?

Who pays if a neighbor's unit has a problem that causes damage to my unit? This is often resolved on a case-by-case basis because there can be a lot of variables. But you want to understand, in general, how your insurance company will handle it. If you let your neighbor know that he might well be responsible for damage in your unit if he or she does not prepare properly, it might spur some action.

How good is my condo association's policy? You will likely have to go to an officer in your condo association to get a copy of the association's policy. Consider sending it to your agent so you know where you stand. You want to know what the association will cover in order to be sure your policy covers the rest.

Have Your Proof

The more proof you have of what you own and how much it's worth, the easier a time you'll have getting reimbursed. These steps can help cut through the red tape:

Take pictures or videos of your house or apartment. A picture can save months of aggravation and a lot of money. Take pictures or videos of everything. Put the "evidence" in a safe, dry place.

Keep your receipts. Keep your receipts in three files: Big Things (furniture, electronics, etc.), Little Things (clothes, household goods, etc.), and Improvements (new kitchen cabinets, light fixtures, etc.).

Make a home inventory. Make a list of all of the big items in your house including serial and model numbers (if applicable).

Keep the list with your important papers. Special computer software is available for this task.

Keep your owner's manuals. Having the owner's manuals for big-ticket items can speed the claim process.

Boat insurance. Boats have their own insurance, and boat insurance has its own set of rules. Just like houses or apartments, however, you should contact your insurance agent to understand the requirements and limits of your policy. Take detailed pictures or video of your boat, inside and out. And keep proof of any add-on items that might be lost or damaged in a hurricane.

When to Buy Insurance

You cannot wait for a storm to approach your area to get insurance. Companies set limits after which they won't issue or change a policy. Flood insurance has a five-day waiting period.

7

Right Before the Storm

Before a Hurricane Watch

The Cone Is Pointing Toward You

The 3-Day Cone is aiming at your part of the coast. It's too soon for a hurricane watch, but people in your neighborhood are getting a little anxious. Remember, even when the cone is pointing right at you, *the odds are that you won't get a direct hit.* But . . . it's a great time to take some early steps to avoid the crush if the storm keeps coming.

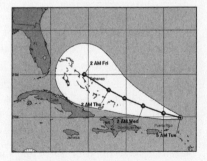

The 3-Day Cone is a good gauge for when you should start early preparations. Don't worry about the 5-Day Cone.

- If you haven't gone through the process of making a family hurricane plan, DO IT NOW. See page 203.
- Read through the following pages so you'll be ready if a hurricane watch is issued in the next couple of days.

- Stock your freezer with water *now*. That way it will have plenty of time to freeze. See "Planning for Drinking Water," page 206.
- Go over the "Shopping Lists" at the back of the book. This is a great time to get hurricane supplies. The stores are normally not crowded and the shelves are full.

Hurricane veterans will tell you that your hurricane-preparation life will be much more pleasant if you take advantage of this time before all of the last-minute preparers hit the stores.

During a Hurricane Watch

Your Plan of Attack

It's time to execute the first part of your family hurricane plan. If you've done things right, everybody knows what they are supposed to do. If you didn't make a plan, go back to page 203 and make some decisions. Then come back here.

> IMPORTANT NOTE: If you are evacuating, you must execute your hurricane plan under a hurricane *watch*. The evacuation is likely to begin when the hurricane *warning* is issued.

Whether you are evacuating or staying at home, you will need to take many of the same steps now that the hurricane threat is real. But there's more time pressure if you're leaving home. Hopefully you already have most of the items on the shopping lists beginning on page 325. But for most people, a trip to the store is still necessary. So your plan of attack is:

 1. Finish up your hurricane shopping lists as soon as possible.

2. Get cash. ATMs won't work after the storm.

3. Get gas. A fill-up now will save you a lot of aggravation later.

And a few small things:

4. Keep your cell phone charged.

5. Run the dishwasher. You may not have a chance for a while.

6. Do the laundry. No power means no clean clothes.

7. Charge your electric razor.

8A. **If you are evacuating**, start setting aside the things you'll take with you. See "Evacuating Checklist" on page 307. Then continue to "During a Hurricane Warning," on the following page.

8B. **If you are staying home**, your next step is to be ready for a hurricane *warning*. For most people who have plywood or storm panels, that means checking them over, being sure you have all of the bolts, and perhaps laying them out. Otherwise, just look though the lists of things you'll be doing when the warning is issued. Be sure you're ready to move quickly.

When a hurricane *warning* is issued, continue on.

9. It is time to fully prepare. Follow the steps on page 314.

During a Hurricane Warning

(OR DURING A HURRICANE WATCH IF YOU ARE EVACUATING)

Executing Your Hurricane Plan

Where is my family going to ride out the storm?

This is the most important question: Hopefully you've thought this through already. If you know you're evacuating, skip to the next section, "Where Am I Going to Get Drinking Water?"

The decision to evacuate is often not easy. In general, you *should leave* home if:

- You don't have confidence in the strength of your home.
- You don't have shutters for the windows and doors.
- You live in an evacuation zone.
- You live in a high-rise and the building's systems are going to be shut off.

To help you make the right decision and understand the ramifications of your decision, go to "Evacuation Decision Making," page 207. Regardless of whether you stay or go, you still have to continue on with preparations. If you are evacuating, however, put aside what you'll be taking with you as you prepare your house or apartment. As a guide, use the "Evacuating Checklist," page 307.

Where am I going to get drinking water?

Whether you go or stay, you're going to need a supply of drinking water during and after the storm. If you haven't done it already, check page 206 for information about your options for water.

If you haven't yet filled your freezer with bags of water, do it now. Fill (90 percent full) jugs or, better yet, collapsible water containers with tap water and put them in the freezer. Move the food to the bottom and then fill up as much of the freezer as possible. You can use Ziploc bags as well.

Do NOT turn the freezer to a colder setting. (I know you've heard the opposite—trust me.)

Do NOT place any bags of ice you bought at the store in the freezer. They often leak when the ice melts. Use this ice along with cold packs in your coolers to keep things cool during the storm.

Buy jugs of water or fill jugs and bottles with tap water so that, along with the water you have in the freezer, you'll have at least five gallons of water for every person in your family.

Store your water in a dark place where the sun can't heat the bottles.

If you're evacuating, try to take three gallons per person with you, but leave plenty at home for when you return.

How am I going to protect my property?

Here is where you're going to do the most work. Obviously, the better you prepared ahead of time, the easier it will be now. Hopefully, you have a feeling for how long it is going to take to prepare. Give yourself plenty of time. Doing this right could make all the difference in the world.

The most important thing you can do to safeguard your property, in general, is to have good shutters for your windows and doors. So let's start there.

Protecting Your House or Apartment/Condo

Windows shutters. If you have accordion shutters, roll-downs, Bahama shutters, colonials, or other shutters that you can close quickly, you can normally wait a while to do it. When it starts getting breezy, however, your shutters should be closed.

If you have storm panels or plywood panels, you're going to have to put them up much sooner. You must have them up before there is a significant breeze. Panels can easily get caught in the wind and you could get hurt.

Shutter Decision Time

The decision time on shutters is tricky. Emergency managers and the National Weather Service could be much more helpful in providing guidance on when the winds will be high enough that putting up panels becomes dangerous. In most storms situations, you should have six to eight hours after the hurricane warning is issued to put up panel shutters safely. That's not always the case, however, so the best advice is to do it early.

Shutter all windows, entry doors, and garage doors, except you *must* leave two entry doors available for a quick exit in case of a fire.

Entry doors. The best entry doors to leave unshuttered are single doors that swing out. The weakest entry doors are normally double doors. They need to be shuttered and have extra bolts added. If all of the entry doors swing in, you'll need to use your best judgment on which ones are strongest, add extra slide-bolts to them, and leave them unshuttered. See page 239 for details.

Garage doors. Many garage doors are very weak, and may be the most vulnerable spot in your house. The first step is to shutter them, if they are not certified as hurricane-resistant on their own. The second step is to strengthen the weak doors. See page 241 for some ideas.

Sliding glass doors. If they are not modern doors designed to resist hurricanes, your sliding glass doors can be a weak spot in your home or apartment. It's important that you lock them and hold them tight in their track. The easiest way to do this is with two-by-fours. See page 242. Most sliding doors also need to be shuttered.

Protecting Your Valuables

There are four levels of protection you should have for your possessions:

1. The most secure, mostly for papers and small items—a *safe-deposit box* at the bank.
2. The next most secure, mostly for papers and small items—a *fireproof and waterproof lockbox* at home.
3. The next most secure for papers and larger items—a *plastic storage box* covered with heavy-duty trash bags.
4. The next most secure for large items—*heavy-duty plastic trash bag* (contractor bags are best).

Ideally you'll use all of the protection options for the appropriate items.

Safe-deposit box. If you have time, take your most valuable papers to the bank and put them in a safe-deposit box. That's where they belong anyway. These might include marriage certificates, divorce decrees, birth certificates, immigration records, wills, trust documents, titles and deeds, and anything you are not going to need right away.

NOTE: Items in a safe-deposit box are normally *not* covered by your homeowners insurance. Seal valuable papers in plastic bags in case the bank is flooded. Check with your insurance agent about insuring other items.

Fireproof and waterproof lockbox. Photos or video of your home or apartment and valuable property, irreplaceable photos and pictures, computer backup disks, jewelry, insurance policies, immunization records, Social Security cards, passports, copies of immigration records, bank statements, investment records, credit card records, income tax returns.

Plastic storage box. Put anything else that is valuable (and not too big) in plastic storage boxes.

Heavy-duty plastic trash bags. Put the lockbox and the storage boxes in trash bags and tape them closed. Put anything big—pictures, electronics, valuable lamps, and so forth—in trash bags. To be sure, use two trash bags. Aim the openings in opposite directions. Tape them closed. Take the lockbox and perhaps one storage box with you if you evacuate.

In-Home Storage Locations

A shelf in an interior closet. Put as many boxes as you can on high shelves. Under a stairway is an excellent location.

A bathtub that you have not filled with water (see below). The structure of the tub will provide some protection.

The dishwasher. A dishwasher with a locked door is, obviously, watertight and normally strengthened by the surrounding counter. *This is also a good alternative if you've waited until the last minute to protect your photos and other valuable papers.*

Plastic sheeting. Wrap large TVs in plastic sheeting. Put plastic sheets over major pieces of furniture if there is a risk of a roof leak. Tape everything as tight as you can.

Furniture and rugs. Move furniture away from the windows. The windows can leak in strong winds, even with shutters. Take rugs off the floor.

Computer. Back up your important computer files and put the backup drive or disks with your important papers. Wrap your computer in a plastic bag or two and put it above the floor level. See page 261 for more computer tips.

Security system. You're probably going to have to disconnect the battery in your security system or the beeping will drive you crazy.

Preparations Inside Your House or Apartment

Sanitary water—use the tub. Line your tub with a plastic sheet and fill the tub with water. This water is for sanitary purposes only, not for drinking.

Mattress. Identify one or two mattresses that you can easily use to protect yourself if it should become necessary. King-size mattresses are not good because they are too hard to handle. For most people, full-size is the best.

Prepare electricity—mark your circuit breakers. Go through your breakers and find one that goes to a bedroom outlet. Plug a lamp into that outlet and turn it on. Also, mark a breaker for an outside light (if you have one).

If you are evacuating, the safest thing to do is to turn off all of your breakers. When you return, first turn on the single breaker that goes to the bedroom light. When that light comes back on, you know you have power. This method isolates anything sensitive in your house from surges that can come down the line when the power is first restored.

If your outside light survives the storm, you'll turn on that breaker as well so that the power company will know your electricity is on. They often work at night.

If you are staying home, when the power flickers or goes out, turn off all the breakers except the one that goes to the bedroom light.

Natural gas. Find your natural gas cutoff valve. Be sure you have a wrench that you can use to close the valve. If you are evacuating because of a flood threat, close the valve before you leave home. Otherwise, unless you are ordered to do so, do *not* turn off the gas coming into your house. You will need it after the storm. Gas service needs to be reinitiated by a professional, and an expert may not be available for some time after a hurricane.

Propane tanks. If there is a threat of flooding in your area, close the valve on any propane gas tanks. Numerous fires were caused after Hurricane Katrina when propane tanks floated away in the flood.

Water valve. Locate your water shutoff valve. Be sure you have the kind of wrench that will fit the valve, if you need one. In general, however, it is not necessary to turn off your water unless you are ordered to do so.

Refrigerator. Put a thermometer in the refrigerator section so you can gauge the temperature after the storm.

If you are evacuating, and have decided to turn off your electricity as recommended, remove anything from the refrigerator that will melt or spoil. You don't know how long the power will be out. Put any food that is in a leakproof container as close to the bottom of the refrigerator or freezer as possible. Fill the rest of the freezer with jugs or bags of water as indicated above.

If you elect to leave your electricity on and hope for the best, do *not* turn your refrigerator to the coldest setting; that won't make the freezer any colder. Be sure you don't have anything in the freezer that might melt and make a mess. Don't forget to turn off the ice maker and put the ice in plastic bags or get rid of it.

Outside the House

Check the yard. Be sure you have secured everything around the outside of the house. Anything that is movable might be picked by the wind. Try to put especially heavy planters or other cement items as close to the house as possible. Pick up any debris or anything that might blow in the wind. Put it somewhere where it can't get caught in the wind.

Do NOT put anything out at the curb unless you are sure it will be picked up. Normally there is no trash pickup after a hurricane *warning* is issued. Similarly, do NOT trim trees under a *warning*.

Check the roof. Remove turbines, antennas, or anything else on the roof that is vulnerable to the wind. Install the turbine covers or cover the openings with heavy-duty plastic and wrap with duct tape. If you have solar panels, turn off the water to them.

Pool or patio screening. The screen in your porch or pool enclosure catches a significant amount of wind. You will lower the risk that the structure holding the screening will be destroyed if you remove the rubber splines on three sides of the screen panels. Leave the top side connected so the panels may flap in the wind. That way, if you're lucky and the screens survive, you'll be able to use your porch after the storm and have some protection from mosquitoes.

Car(s). Wherever you end up, think about where you're parking your car. Do not park under or near trees. Put the car as close to the house or a building as possible. Try to park on the opposite side of the house from the wind, if you can determine that. See page 274.

When You Think Everything Is Done, Double-Check Yourself With

House Preparation Checklist, page 314

Apartment/Condo Preparation Checklist, page 317

In addition, if you're staying home, you'll need to prepare a safe spot in your house or apartment. Go to:

Staying-in-a-House Checklist, page 320

Staying-in-an-Apartment/Condo Checklist, page 322

8

During the Storm

What to Expect

The storm is said to *begin* when the winds reach about 40 mph. At that wind speed bad things start happening: some tree limbs snap, some signs blow down, debris starts blowing around. Many emergency vehicles will no longer respond to calls. It's time to hunker down.

> Prepare your family for the possibility they might have to stay put for an extended period of time: from a couple of hours to twelve hours or more during a large, slow-moving hurricane.

How It Sounds

The wind will whistle and howl at the same time. You may hear pieces of debris hitting the house or the shutters. In a house with an attic crawl space, the hatch cover may bounce up and down in the wind throughout the storm, making an annoying noise if you haven't taken it out or latched it. Shutters can also make a racket shaking in the wind.

How It Feels

If the eye of a hurricane comes over or near you, the atmospheric pressure will drop very quickly. This creates a pressure imbalance in your body, an extreme version of what happens when you go up in a fast elevator. Air and gasses inside your body, which were equalized with the atmospheric pressure before the storm, want to push toward the lower pressure outside. This creates a sensation of pressure pushing on your body. You can't tell whether the pressure is inward or outward.

Not everyone will experience exactly the same sensations. People with arthritis, for example, may feel pain in their joints because the trapped gases push out on enflamed tissue.

Studies have shown that childbirth is *not* induced by the low air pressure, although deliveries go up. Likely the increase is caused by the stress associated with the storm.

Staying Safe

Don't take chances. Follow these safety rules:

When the Winds Reach 40 MPH

You don't have to measure the winds precisely. You'll be able to tell. It will seem quite windy outside. Trees will bend over and you'll see some debris blowing down the street.

- NEVER use candles or have any open flame while the wind is blowing.
- Stay away from unprotected windows. Don't give in to the temptation to see what's happening through an unprotected window.
- If the power blinks off and on, turn off all of the circuit breakers except the one that powers a lamp in an internal part of your house or apartment. If you haven't already, see page 289.

- Put any food you're going to want to eat during the storm in coolers. Then don't open the refrigerator door.
- Put the finishing touches on your safe spot. If you haven't created one, go to page 320 if you're in a house or page 323 if in an apartment.
- If you rely on an elevator to get to your safe spot, you need to go there now. You don't want to be in an elevator when the power goes out.
- If you haven't done it, turn off your burglar alarm and take out the batteries. Otherwise it will start screeching when the power goes out.
- Call your emergency contact to let them know where you are and how you are doing.

When the Winds Are 60 MPH and Above

You'll know it. You'll start hearing the wind whistle and the rain will be horizontal.

- NEVER, EVER, UNDER ANY CIRCUMSTANCES LIGHT A MATCH.
- Stay away from windows and doors, even if they are shuttered.
- Do *not* stay above the third floor in a high-rise.
- Move to your safe room or area and plan to stay there. See page 320 for a house and page 323 for an apartment to be sure you have everything you might need.
- Listen to radio or TV so you know where the storm is and whether the eye might be passing over your location.
- If a window breaks or you hear some other noise, stay put.
- If you become concerned for the integrity of your house, or the roof starts to leak so the ceiling dry wall is getting wet, retreat to a closet or a corner and get under a mattress. Often, the absolutely safest place is in a bathtub under a mattress.
- Be very cautious if the eye passes over your location. It will likely be impossible for you to know how much time you have to make any emergency repairs. The strong wind may return *very* rapidly, perhaps in just a minute or two.

- Stay put until you are sure the wind has died down. The fringes of the storm can fool you. Strong bands may still come though with fairly calm wind in between.
- Use extreme caution. You may not be able to get help if you hurt yourself.

9

After the Storm

What to Do When the Wind Stops

> **Use Extreme Caution!**
> In most hurricanes, more people die after the storm moves by than during the event itself. People use equipment they are not familiar with, traffic signals are out, electric lines are down, and medical services are not functioning normally.

Try to be sure the storm is really over before you venture out. Gusts may still pick up dangerous debris well after the strongest winds have passed. Be sure everyone understands the precautions they need to take.

- **Downed power lines.** NEVER assume that a power line lying on the ground is dead. There have been numerous sad stories of people walking dogs or just venturing outside and being killed by an energized wire hidden in a puddle.
- **Power out.** Do NOT call to report that your power is out. DO call to report a downed line.

- **Gas leaks.** If you smell gas, immediately turn off the main gas valve.

- **Flooded roads.** Prior to Hurricane Katrina more people had died after hurricanes because of inland flooding, than in the hurricane storm surge. Many deaths are caused when people drive on flooded roads. Do NOT assume you know the depth of the water. Do NOT assume that the ground is solid under a flooded roadway. Anything could have happened during the storm.

- **Children.** Keep a close eye on children until you can assess the situation. Adventurous children are often the victims of post-hurricane accidents.

- **Generator.** If you have a generator, follow the rules. Locate it well away from the house—never on a porch, in a garage, or under an overhang. Use heavy-duty extension cords. Install carbon-monoxide detectors. See page 255 for more information.

- **Pets.** Do NOT let pets roam around outdoors. Their curiosity could be fatal to them.

- **Call out.** If everybody is okay and you have a phone that works, call your contact person to let them know your situation. Phones are less likely to work later. Then, stay off the phone except in an emergency.

- **If you evacuated.** You may be barred from going home for some time. Do NOT get angry with the law enforcement people who are keeping you away from your house. They will let you go back as soon as they do a complete assessment and are sure it's safe.

- **Make an assessment.** Assess your situation carefully. Look for hazards around the house. If you had significant flooding or structural damage, you shouldn't stay at home. You'll need an evaluation from a professional to determine what needs to be done to make it safe to return.

- **Pictures.** Take as many pictures as you can for the insurance company. If you have a film camera, take two shots of everything.

- **Power.** If you had any flooding, turn off the main power switch or all of the breakers. Check the meter and the pipes where the lines come into the house for damage. If repairs are required, you'll need to call an electrician. If there are no obvious problems and your neighbor's power is restored while yours is still off, call the power company.

- **Breakers.** If you did not have flooding, leave all the breakers *off* except the one you marked for the inside lamp and the one for the outside light (if the fixture survived the storm). *When it's time to turn on all of the breakers, have a fire extinguisher ready.*

- **Debris.** Do *not* pile debris around power poles, electrical boxes, cable boxes, fences, or downed power lines. It will only slow down debris pickup and utility repairs.

- **Make temporary repairs.** Do what you can to keep any further damage from happening, but don't do anything unsafe. Don't make any permanent repairs until the insurance adjuster has seen the property. Take pictures of what you do so there are no questions later.

- **Mark your property.** If your house and neighborhood have suffered major damage making it difficult for an insurance adjuster to find you, paint your name, address, and insurance company on the front of the house.

Collecting on Insurance

- **Make a call.** Call your agent or your insurance company as soon as possible. Companies sometimes set up special numbers. If you can get access to the Internet, go to the company's Web site for contact information. Remember, damage caused by rising water will be covered under your flood insurance policy. Damage caused by the wind will be covered either under your homeowner's or wind policy, depending on where you live.

- **Keep every receipt.** Receipts and pictures are often the grease that speeds the process of claims being paid. Keep receipts for temporary housing, temporary repairs, restoration expenses, and car repairs.

- **Pictures again.** Another reminder to take lots of pictures. The trail of photos from before the storm, immediately after, and then after temporary repairs should speed things along.
- **Document your loss.** Make an inventory of your losses. If you prepared well before the storm, you'll have owner's manuals and serial numbers for some items. But, even if you don't, a good list with pictures will go a long way toward substantiating your claim.
- **Living expenses.** Many policies will give you an advance on your temporary living expenses. Ask your agent.
- **Your deductible.** Ask your agent about your deductible. The hurricane deductible in many states is a *percentage* of the value of the structure (not including the land), instead of a fixed amount. It is usually a significant amount of money.
- **Debris removal.** Some policies will pay some debris removal costs. Also, your municipality may handle most debris removal with funding from FEMA. Check with your agent.
- **Trees and shrubbery.** Insurance, in general, won't pay for damage to any landscaping. Normally your policy will pay (up to a limit) to remove trees that have fallen on your house. If a tree from your yard damages your neighbor's house, it's usually *your* insurance that will pay, up to a point. You will likely only get partial payment for removing the tree from the property.
- **Time limit.** You can't wait forever to file your insurance claim. The limits vary, depending on state laws and policy provisions. In general, you are best off filing as soon as possible.

Apartments and Condos Only

Who pays? In general, your landlord or condo association's insurance will *not* pay for your personal possessions, even if the roof leaked or there was a building problem that was not your fault. You need to have renter's insurance for your things.

Some condos may pay to have your unit restored to the way it was when it was new, not counting any renovations or improvements you have made. But most often the association's insurance will only pay to rebuild up to the bare walls.

Temporary living expenses. Most renters' policies have a provision for paying any living expenses over what you would normally pay. Check with your agent.

Your Car

Flooding or vandalism. If you have *comprehensive* insurance as part of your auto policy, your car is covered if it is flooded or otherwise damaged. If you had personal items in the trunk, they would normally be covered under your *homeowner's* policy.

Picking a Contractor

Look out for scammers. Unfortunately, hurricanes bring scammers out of the woodwork. Assume that any contractor who pressures you to sign immediately or demands a large payment up front is not honest.

Temporary-repair payments. Beware of a contractor who shows up and offers to do temporary repairs for a significant amount of money. Don't do it.

Get the license number. Only use licensed contractors. Get their license numbers and references and check them out before paying any money.

If the Insurance Company Doesn't Pay

Don't give up. It you don't get a fair settlement from the insurance company, call your agent and find out how to present your case to the head of the claims department. Your second stop is your state insurance commission.

Public adjusters. You can hire a public adjuster to help you make your case to your insurance company. Call your state insurance commission for information, including fee regulations, on public adjusters.

Eating After the Storm

Refrigerator. As a rough rule of thumb, if the power is out for less than four hours and you haven't opened the refrigerator, it probably stayed cold. When the lights come back on, open the door just to check. If the temperature inside is less than 40°F you are in good shape.

If the refrigerator got above 40°F (did you buy that fridge thermometer on the shopping list?) for more than two hours, you have to throw away all of the perishable food. This includes proteins and dairy—meat or fish, cooked pasta, anything with egg, stews, mayonnaise, cream-filled pastries. Don't taste or eat anything that looks or smells funny. *When in doubt, throw it out.* Keep the refrigerator closed as much as possible.

Freezer. A full freezer should last about two days without power if you don't open the door.

Flood food. Throw out any food that got wet unless it's in an airtight container.

Moldy food. Throw it out. Scraping off the mold isn't good enough. Bad stuff may have gotten into the food.

Frozen food with ice crystals. If the food is still frozen but is covered with ice crystals, it's still good. You can eat it or refreeze it. If it is thawed but has not been above 40°F for two hours you can cook it and eat it.

Cooking. NEVER use a grill or camping stove inside the house, garage, or porch. Always have good ventilation.

Drinking Water After the Storm

Self-bottled water. Drink the tap water you bottled before the storm first. It will be good for about a week at room temperature. The water that was frozen and stayed cool should be good for another week.

> ### My Water Tastes Flat!
> Any stored or boiled water will taste better if you pour it back and forth between two clean containers before drinking it. This will put oxygen from the air back into the water and remove the "flat" taste.

Water heater. It's likely you have a good source of drinking water in the tank of your water heater. Turn off the breaker to the heater, close the intake valve, and let the water cool. Then you can drink it.

Tap water. When the electricity goes out or falling trees uproot water mains, the system depressurizes and contaminants get into the water. The pipes have to be repaired and then repressurized and flushed before the water is safe to drink. If you have no other choice, there are things you can do to make the water safe.

Purifying Water

Boil tap water. This is the best way to purify water, but obviously you have to have a good source of heat.

- Strain the water through a clean cloth to remove sand or other large impurities.
- Bring the water to a rolling boil for about a minute.
- When it's cool, pour the water back and forth between two clean containers.

Use bleach with tap water. Household bleach can be used to treat tap water. Use only nonscented bleach with no additives in a *new* bottle. Check the label. The product should contain about 6 percent sodium hypochlorite as the only active ingredient.

- The purification process works best with warm—or at least room-temperature—water.

- Add sixteen drops or one-eighth teaspoon of bleach to a gallon of tap water.
- Stir and let it stand for thirty minutes.
- The water should smell slightly of chlorine. If it doesn't, don't drink it. Get water elsewhere.

Use water-treatment tablets. Purification tablets that have sodium hypochlorite as the active ingredient are recommended. Iodine-based tablets require the water to be kept in a dark bottle and may be a problem for some people (those with thyroid problems, for example). Follow the directions carefully on the package. Don't use any tablets after their expiration date (they don't last forever) or if the container has been open for some time.

Saving Your Stuff

Water got in the house. Try to get the sand or mud out of the house while it's still damp. Use freshwater to rinse anything that came in contact with floodwater.

Saving furniture. Take any furniture that got wet outside, but do not place it in direct sunlight. Remove drawers, but don't force them. They may come out better when the wood dries. Remove any dirt or flood coating from the wood.

Rugs and carpeting. Wash rugs well with freshwater and then put them in the sun to dry. Move them around occasionally so you don't get mildew on the bottom side. It will take quite some time for rugs to dry completely.

Valuable papers and other items. Your only shot is to soak them in a shallow pan of clean water, then gently move them to a table to dry. If you have a piece of screening, it might help you move the papers without having to handle them while they are wet.

Getting Help from FEMA

Call! If you have suffered a loss from a hurricane, call FEMA. They will reimburse you for an amazing number of things. Keep the number handy.

1-800-621-FEMA

> You may have the creeps about calling FEMA, and you may not agree with their policies for doling out money, but if you pay taxes, you have a right to benefits under the established rules.

Operation Blue Roof. FEMA has a good program that can help some people who suffer roof damage. There are a bunch of rules, though. Not everyone qualifies.

- Your roof has to be less than 50 percent damaged, and still be structurally sound.
- You must have a right of way so that contractors can come on your property.
- Government contractors will do the work and will get to you as soon as possible.
- In some cases, you can get free tarps that you can put on.

Call 1-888-ROOF-BLU for information on how to sign up.

10

Preparation Checklists

Evacuating Checklist

If you've decided to leave your home to ride out the storm in a safer place, use this checklist to be sure you have the critical things you might need. Have plastic storage boxes, heavy-duty plastic bags, and duct tape ready before you start packing. Here are your most-important first steps:

 Fill your gas tank. Don't wait until you're on the road with everyone else.

 Charge your mobile phone. And be sure you have a car-charger cord for your phone. There may not be electricity where you're going.

 Take cash, checkbook, and credit cards. And get at least one roll of quarters in case cell phones don't work and you need to use a pay phone.

 Back up your computer. Do one final backup on an external drive, CD, or DVD and bring the backup with you. Do it even if you are bringing your computer as well.

☑ **Get a map of your route.** Be sure you have a map if there's any chance you'll end up on roads that aren't familiar. You may need to detour.

☑ **Google Earth Map showing gas stations and restaurants.** If you know how to work Google Earth Map on your computer, plot out the route you plan to take showing the gas stations along the way and on nearby roads. You can do the same for restaurants. Print out the maps.

☑ **Call your family contact.** Before you get on the road, call your family contact out of town and let them know you're leaving. See "Family Communications Plan," page 204.

☑ **Take important papers, files, pictures, and heirlooms with you.**

For the Lockbox—In the Trunk

Put all of your papers in plastic Ziploc bags. Two-gallon bags will hold most papers and files perfectly. Put the most important and irreplaceable papers, photos, mementoes, and computer backups in a waterproof and fireproof lockbox, if you have one.

Put everything else in plastic bags, then inside a plastic storage box. Duct tape the box closed to keep the top from coming loose. Put both the waterproof and fireproof box and the taped box in a heavy-duty plastic bag, then seal the bag with duct tape.

☑ **Insurance papers.** Bring a copy of important insurance papers and contact numbers for the companies.

☑ **Photos or videotapes of your home and property.** Bring photos or videotapes of all of your household possessions in case you need to collect on insurance. If you are making hard copies, make at least two of each item. Make more than one videotape, that way you can keep a copy and send one to the insurance company. If you have a digital camera, e-mail pictures of your property to a relative out of town.

Inventory with model numbers and serial numbers. Do an inventory of your home. Copy down all of the model and serial numbers of your appliances and electronics and bring the paper with you.

Owner's manuals. Bring the ones you have on hand to prove you owned the equipment.

 Computer backup. Bring the backup drive, CD, or DVD, or USB drive.

 List of bills coming due. Be sure you don't get behind in your bills because you are evacuating.

 Wills, deeds, leases, and important legal papers. Bring a copy of any critical legal papers and documents like powers of attorney, living wills, trust documents, and so forth.

Birth, marriage, adoption, and death certificates. These should really be in a sealed plastic bag in the safe-deposit box.

 Tax returns. Bring your tax returns for the past few years. The first two pages are the most important.

 Social Security card. Just to keep it safe.

Stocks and bonds. Bring them if you have any at home. They should, of course, be sealed in a plastic bag in a safe-deposit box.

Titles and ownership documents. Bring any documents that prove you own something. These should also be sealed in a plastic bag in a safe-deposit box.

 Benefits documents from your employer. In case you have health problems while you're gone.

Important keys. Put keys to your safe-deposit box, safes, relatives' homes, your workplace, and so on in the lockbox.

 Photos or family pictures. To the extent you can, pack irreplaceable photos and pictures in their own plastic box inside the larger box.

 Jewelry or family heirlooms. Same as the photos. Put what you can in a small plastic box inside the larger box.

Hurricane Kit—Inside the Car

Divide your hurricane kit into two parts. In the trunk, put the things you might need where you're going. Put the things in the car that you'll need on the road. Put all of the papers you're taking with you in a plastic bag to keep them separate.

 Emergency phone numbers. Make a list of the phone numbers of family members, doctors, your pharmacy, financial advisors, friends who could check on your house, insurance agents, and so on.

 Driver's license or ID card. Keep the original in your wallet and put a copy in the box.

 Passport. Keep a copy in your wallet and the original in the box.

 Proof of residence. Utility bills are a good proof of residence in case you need one to get back home.

 Prescriptions and medical papers. Don't forget paperwork about your glasses in case you need new ones. Bring children's immunization papers in case there are any questions.

 Flashlights. Have at least one big one that makes lots of light plus one for each person.

Mobile phone charger. Keep the phone charged while you're driving.

 Food and drinks. Breakfast bars are great in the car. Single-serving juices are best, especially if you have a cooler.

 Water. Bring a gallon or two of water inside the car.

 Medicine. Put some medicine in the car and the rest in the trunk.

 Games. The traffic may be terrible, so have something for the kids to do.

 Map. Be ready to find a detour if it becomes necessary.

 Reading glasses. If you need them to see the fine print on the map.

 Checkbook. Just in case.

 Google Earth Maps. If you printed them showing your route with gas stations and restaurants.

Hurricane Kit—In the Trunk

This will be a box or bag of things you'll need when you get where you're going.

 Batteries. Be sure you have batteries for *everything* that you might need. Look around again. It's easy to forget something.

 Clothes. Bring enough clothes for several days, but don't bring your whole wardrobe. Jeans, shorts, and loose-fitting clothes that you can wear more than once are best.

 Toiletries. Bring everything you'd take on a trip. Charge your electric razor.

Bedding. Bring a pillow, blanket, and a sheet for each person. Bring sleeping bags or air mattresses if you have them.

 Paper products. Bring a couple of rolls of paper towels and toilet paper.

 Tool kit. Bring a small tool kit and work clothes.

 Fix-A-Flat. Keep a few cans of Fix-A-Flat or a comparable product in your trunk. The last thing you need is a flat tire, and posthurricane debris increases the chance of a problem.

 Food kit. Bring as much of your hurricane food as you can fit in a large plastic box. You don't know how available food will be where you're going.

 Water. Carry at least three gallons of water per person in the car. Water is the most difficult to pack. You may want to use collapsible water containers. If you don't fill them all the way, you can mold them to fit in odd-shaped spots in the trunk.

 Medicine. Bring any medicine you might need and any special medical equipment.

 Mobile phone charger. Don't forget your plug-in mobile phone charger.

 Battery-operated radio, TV, lights. If you have room for them, bring any radios, TVs, and lanterns or flashlights. You never know if the power will stay on where you're going.

 Call your emergency contact person to let them know where you're going.

Plan ahead. Leave early!

Shelter Checklist

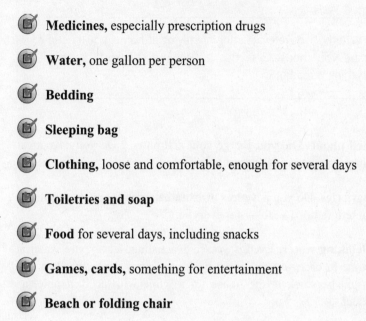

Medicines, especially prescription drugs

Water, one gallon per person

Bedding

Sleeping bag

Clothing, loose and comfortable, enough for several days

Toiletries and soap

Food for several days, including snacks

Games, cards, something for entertainment

Beach or folding chair

House Preparation Checklist

> Whether you're evacuating or staying at home, you'll need to prepare your house. Use this last-minute checklist to be sure everything is ready.

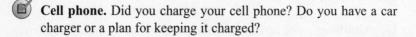 **Cell phone.** Did you charge your cell phone? Do you have a car charger or a plan for keeping it charged?

Batteries. Do you have lots of extra batteries for every flashlight, lantern radio, TV, toy—for everything?

Drinking water. Even if you're evacuating, leave extra water at home in case you can come back soon after the storm. You still might be stuck in the house for a while with no drinkable tap water.

Whether you stay or go, fill your freezer with jugs or plastic bags of water well before the storm so they can freeze. They will keep your refrigerator colder longer and supply your after-storm drinking water. If you're staying, put extra water in your coolers for use during the storm.

Sanitary water—use the tub. As long as the water and sewer systems are working, you will be able to use your toilet. If, however, the water system fails, you'll need your own water supply to flush. If the sewer system fails, you'll have to use plastic bags for a toilet.

The easiest place to store water for sanitary purposes is your bathtub. Close the drain, line the tub with a plastic sheet, and fill it with water. If you're evacuating, just leave it there until you come back.

To flush, use a bucket to fill up the toilet's tank and flush normally.

Shutters. All windows, entry doors, and garage doors should be shuttered (except as below). Check to be sure all of the shutters are tight to the house and any latches are properly set. See page 221.

 Doors. Have you left two entry doors unshuttered so you can escape if there's a fire? Have you secured the weak doors: garage doors, double doors, sliding glass doors? See page 238.

 Safe room or area. If you are going to ride out the storm at home, you're going to have to prepare a safe room or area in the house. See "Staying-in-a-House Checklist," page 320.

 Protect valuables. Have you put everything valuable in secure boxes high off the ground? Have you covered everything you can with plastic? Have you moved furniture away from the windows?

 Check the yard. Have you picked up everything that might blow in the wind?

 Check the gutters. Do what you can to be sure water can flow away from your house. Work with your neighbors to clean debris out of gutters in the street.

Check the roof. Is anything left on the roof that could come off and blow around?

Check the porch screens. Did you take out the screens or let them flap free so the wind will be able to blow through?

Car(s). Have you thought about where you will park your car? Obviously a garage is best. But if you have to leave it outside, get it away from the trees and park it close to the house, on the side away from the wind if possible.

 Electricity. Did you mark your circuit breakers so you know which ones to turn off and which one to leave on?

> IMPORTANT: If you are cooking on an electric stove when the power goes out, turn the stove off at once, before you forget.

 Natural gas. Did you find the cutoff valve for the house in case you need to close it?

 LP/propane gas. If you are in an area likely to flood, did you close the valves on your propane tanks? Don't forget the grill.

 Water. Did you find the cutoff valve for the house in case you need it?

 Computer. Did you back up your important computer files and put the backup drive or disks with your important papers?

 Electronics. Did you put audio components and other electronics in plastic bags and put them above the floor level?

 Refrigerator. Did you take out any food that will melt or spoil if the power goes out? Did you put the thermometer in the refrigerator section?

 Electric razor. Did you charge your electric razor?

 Anything else valuable. Use common sense. There is a chance that water will come into the house. Cover big items and put smaller items in double plastic bags. Aim the openings of the two bags in different directions and tape them up to make them as watertight as possible.

Apartment/Condo Preparation Checklist

Whether you're evacuating, moving to a lower floor, or staying in your apartment or condo, you'll need to prepare for the storm. Use this last-minute checklist to be sure everything is ready.

Cell phone. Did you charge your cell phone? Do you have a car charger or a plan for keeping it charged?

Batteries. Do you have extra batteries for every flashlight, lantern radio, TV, toy—for *everything*?

Drinking water. Even if you're evacuating, leave extra water at in your apartment in case you can come back soon after the storm. You still might be stuck there for a while with no drinkable or working tap water.

Whether you stay or go, fill your freezer with jugs or plastic bags of water well before the storm so they can freeze. They will keep your refrigerator colder longer and supply your poststorm drinking water. If you're staying, put extra water in your coolers for use during the storm.

Sanitary water—use the tub. Tall buildings have to pump the water to the upper floors, so if the power goes out, there is a chance the water will go out as well. That means you won't be able to flush the toilet. So, in high rises, it's doubly important to have water saved for sanitary after-storm purposes. The easiest place to store this water is your bathtub. Close the drain, line the tub with a plastic sheet, and fill it with water. If you're evacuating, just leave it there until you come back.

To flush, use a bucket or large pot to fill up the toilet's tank and flush normally.

Shutters. All windows and outside doors should have hurricane protection. (See page 221.) Check to be sure all of the shutters are closed and secured.

Entry doors. If you have entry doors that go to the outside, are they strong as they can be? (See page 239.) If your door opens to an interior hallway, be sure it is securely locked.

Sliding glass doors. Are your sliding doors locked and jammed closed with wood or some other means? See page 243.

Safe room or area. If you are going to ride out the storm in your building, you're going to have to prepare a safe spot in your apartment or on a lower floor. See the "Staying in an Apartment/Condo Checklist," page 322.

Protect valuables. Have you put everything valuable in secure boxes high off the floor? Have you covered everything you can with plastic? Have you moved furniture away from the windows?

Car(s). Have you thought about where you will park your car? Obviously a garage is best. But do not leave your car in a below-ground parking garage. If you have to leave it outside, park it away from the trees and put it as close to the building as you can, on the side away from the wind if possible. See page 247.

Electricity. Did you mark your circuit breakers so you know which ones to turn off and which one to leave on?

> If you are cooking on an electric stove when the power goes out, turn the stove off at once, before you forget.

Natural gas. Did you find the cutoff valve for your unit?

Water. Did you find the cutoff valve for your unit?

Computer. Did you back up your important computer files and put the backup drive or disks with your important papers?

 Electronics. Did you put audio components and other electronics in plastic bags and put them above the floor level?

Refrigerator. Did you take out any food that will melt or spoil if the power goes out?

Electric razor. Did you charge your electric razor?

Anything else valuable. Use common sense. There is a chance that water will come into the house. Cover big items and put smaller items in double plastic bags. Aim the openings of the two bags in different directions and tape them up to make them as watertight as possible.

Staying-in-a-House Checklist

Under most circumstances, you should stay in your house during a hurricane IF AND ONLY IF:

- You have confidence in the strength of the construction.
- You have shutters for the all of the windows or they have impact glass. See page 221.
- The doors have been shuttered with the latches strengthened or are strong enough not to require shutters. (See page 238.)
- You are outside the evacuation zone.

If any one of these criteria is NOT met, secure your house as best you can and find safe shelter elsewhere. (See page 211.)

Staying at home safely and as comfortably as possible requires taking the necessary steps *well before* the wind starts blowing. In the end, the work may not have been necessary—the storm may weaken or change coarse at the last moment—but if the National Hurricane Center determines that the threat is high, there is no alternative other than to fully prepare.

Double-check what you've done with the "House Preparation Checklist" (page 314).

Prepare a Safe Room or Area

Prepare a room, closet, or hallway in the center of your house as your safe spot, where you'll ride out the worst of the storm. This area should not have windows and should not be very large. Small areas are safer.

Make your safe spot comfortable. You may be there for some time, from a couple of hours to more than twelve hours in a large, slow-moving hurricane.

Take the following to your safe spot:

 Coolers. A couple of coolers of food and drinks.

 Water. A few gallons of water.

 Light. Lanterns and flashlights.

 Radio/TV. Portable radio or TV.

 Batteries. Lots of batteries.

 Plastic. Plastic sheeting in case of a roof leak.

 Cushions. Pillows, cushions, or folding chairs to sit on.

 Bedding. Bedding in case you are there through the night.

 Games. Games, cards, something to do, especially for the kids.

 Telephone. Connect your *corded* (not cordless) telephone to a long wire and run it to your safe spot so you can answer it from there.

 Cell phone. Bring your cell phone and charger to the safe area. Keep it on charge as long as possible.

 Mattress. Identify a mattress or two that you can easily grab if you need protection during the storm.

 Refrigerator. Do *not* turn your refrigerator to the coldest setting; that won't make the freezer any colder. Get what you need for your coolers before the power goes out or you turn it off, then avoid opening the refrigerator door.

 Call your family contact. Once you're organized—and before the wind starts to blow—call your out-of-town family contact and let them know you're staying home.

Staying-in-an-Apartment/Condo Checklist

Staying at home safely and as comfortably as possible requires taking the necessary steps *well before* the wind starts blowing. In the end, the work may not have been necessary—the storm may weaken or change coarse at the last moment—but if the National Hurricane Center determines that the threat is high, there is no alternative than to fully prepare.

Remember, even if you fully prepare your apartment or condo, you are still vulnerable to damage if your neighbors around you fail to do so.

Once the wind starts blowing, go to your safe spot. DO NOT GO IN OR OUT OF YOUR APARTMENT WHILE THE WIND IS STRONG. The wind pressure could rip the entry door out of your hand, doing damage or hurting you or somebody else.

Under most circumstances, you should stay in your apartment or condo during a hurricane IF AND ONLY IF:

- You live on the third floor or below.
- You have confidence in the strength of the construction.
- You have shutters for the all of the windows, doors, and sliding doors, or they have impact glass. See pages 221–244.
- You are outside the evacuation zone.

If any one of these criterion is NOT met, secure your apartment as best you can and find safe shelter elsewhere. If you are NOT in an evacuation zone and your building is well built, if possible, make arrangements to ride out the storm in a hallway, stairwell, or safe room with no windows on the second or third floor.

Double-check what you've done with the "Apartment/Condo Preparation Checklist" (page 317).

Preparing a Safe Room or Area

If you can stay in your apartment building, you'll need to prepare a safe spot. If you're staying in your own apartment, choose a room, closet, or hallway away from the windows as your place to ride out the worst of the storm. Otherwise, set up your safe area in an interior part of a lower floor.

> Make your safe spot comfortable. You may be there for some time, from a couple of hours to more than twelve hours in a large, slow-moving hurricane.

☑ **Coolers.** A couple of coolers of food and drinks.

☑ **Water.** A few gallons of water.

☑ **Light.** Lanterns and flashlights.

☑ **Radio/TV.** Portable radio or TV.

☑ **Batteries.** Lots of batteries.

☑ **Plastic.** Plastic sheeting in case of a leak.

☑ **Cushions.** Pillows, cushions, or folding chairs to sit on.

☑ **Bedding.** Bedding in case you are there through the night.

☑ **Games.** Games, cards, something to do, especially for the kids.

☑ **Telephone.** If possible, connect your *corded* (not cordless) telephone to a long wire and run it to your safe spot so you can answer it from there.

Cell phone. Bring your cell phone and charger to the safe area. Keep it on charge as long as possible.

 Mattress. Identify a mattress or two that you can easily grab if you need protection during the storm.

 Refrigerator. Do *not* turn your refrigerator to the coldest setting; that won't make the freezer any colder. Get what you need for your coolers before the power goes out or you turn it off, then avoid opening the refrigerator door.

 Call your family contact. Once you're organized—and before the wind starts to blow—call your out-of-town family contact and let them know you're staying in your building.

11

Hurricane Shopping Lists

Grocery/Drugstore Shopping List

> #### How Much Do I Need?
> I recommend that you have at least four days of basic supplies and water for every member of your family except for babies, anyone with a medical condition, and pets. Have at least a month's supply of baby food, pet food, and prescription medicines.

Water—at least 4 gallons per person. You should have a gallon per person per day. But in practice this amount will last longer, especially if it is supplemented with juices and other drinks as well. Remember, you don't need to buy bottled water. If you have containers at home, you can fill them with tap water (see page 206). And don't forget water for your pets.

Prescription medicines. Try to have enough to last a month, but check with the pharmacist to be sure you medications will keep that long.

Batteries. Be sure you have lots of batteries that fit your flashlights, lanterns, portable TV/radio, and so on.

Cheap batteries. Batteries can be bought in bulk at a significant discount.

Supercharge your flashlights. Consider replacing the standard bulbs in your handheld C- and D-cell flashlights with LED bulbs. They will last up to *ten times longer*—over a week with normal use. Check www.HurricaneAlmanac.com for more information.

 Baby food and supplies. Buy enough disposable diapers, formula, and food to last a month.

 Pet food and supplies. Have enough for a month. It will keep.

 Special needs. Think of things you might need in the next month that aren't readily available: hearing aid batteries, dentures, special clothing or medical supplies, special diet foods, crutches.

Ice. Get some bags of ice for your coolers. You'll use them during the storm.

Food

Have enough for four days, and a week if you stretch it.

Leftover Hurricane Food

Conveniently, Thanksgiving comes right at the end of hurricane season. Consider giving most of your leftover hurricane food to a food bank in your area. Just keep enough in case something weird happens late in the year. While it is possible to have a hurricane in December, the odds are high it would not be stronger than category 1.

Bread, crackers, and cookies. And a waterproof container to keep them fresh.

Breakfast bars. Get a variety. They are great hurricane food, and they keep for a long time.

Peanut butter and jelly. It's great food! How can you go wrong?

Cereals. And plastic containers to keep it fresh.

Juices. Get the kind in small boxes that don't need refrigeration.

Milk. Get extended-shelf-life milk like Parmalat in small boxes.

Canned soup. Only if you're planning on cooking.

Prepared food. Nonrefrigerated spaghetti, tuna, chicken, ham, fruit cocktail, and pudding.

Fruit (dried and fresh) and raw vegetables. The forgotten hurricane food. They provide good variety for your poststorm meals and are good for you, too.

Other Stuff

 Antibacterial hand soap or hand cleaner. After the storm you should wash your hands a lot.

Bleach, nonscented. Just in case you need to purify water from the tap.

Paper plates and plastic utensils. Get plenty. They last forever.

Plastic cups. Get plenty of these, too. You can always use them.

Paper towels. Get one of those big packages with eight or twelve rolls.

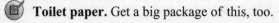 **Toilet paper.** Get a big package of this, too.

 Plastic garbage bags. Get several boxes of big, medium, and small bags with ties. You'll use them before and after the storm. The small bags can be your toilet if the sewer system stops working. You'll also want some extra-heavy-duty construction bags from the home-improvement store.

 Zip-type freezer bags. You'll use them to protect valuables as well as to store water in the freezer. Get large ones, such as the two-pound Ziploc freezer bags, to hold papers and files.

 Aluminum foil. It's useful if you're going to plan on cooking, especially on a grill.

 Oven mitts. If you're planning on heating cans of soup or other food or cooking on a grill.

 Plastic food containers. To keep dry food fresh. Remember, it will be humid in your house after the storm.

 Mosquito repellant. Mosquitoes love to come out after hurricanes.

Disinfectant. Get some Pine-Sol or similar disinfectant for cleaning up. You can also get antibacterial disinfectant at the home-improvement store.

 Candles. Read page 266 before you use any candles. NEVER use candles during the storm. If you do buy them, get only votive candles and glass holders.

 Matches. Get large safety matches and put them in a plastic container so they stay dry.

Butane lighter. Get a couple of the type with a long nozzle.

Sunscreen. If you're going to be outside cleaning up.

 Manual can opener and bottle opener. You need them anyway.

 Personal toiletries kit. Have a kit of the personal things in case you have to leave home. Toothbrush, toothpaste, bottled body soap, shampoo, hair products, and so on.

Safety and Medical Supplies

 Medical information. Write down any special medical information and keep it with you. Or, consider a MedicAlert emblem or HealthKey if you want to ensure that medical personnel know about your condition.

 First aid kit. Get one that's already made up or make your own. Be sure you have bandages, adhesive tape, antibacterial ointment, and alcohol. Put it in a plastic box.

Over-the-counter medicines. Antidiarrhea medicine, pain medicine (aspirin, Tylenol, etc.), antacids, and allergy or cold medicine. Put them in your first aid kit.

Tweezers and scissors and needles. Put them in your first aid kit.

Medicine dropper. In case you have to use bleach to sterilize water.

Home Improvement Store Shopping List

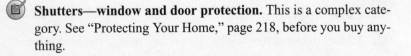

These supplies will normally last you for many seasons, so don't scrimp. If you shop right, your life will be much better after the storm.

 Shutters—window and door protection. This is a complex category. See "Protecting Your Home," page 218, before you buy anything.

 Fire extinguishers. Get two. You should have them anyway. Be sure they are the ABC type.

 Contractor trash bags. These are a heavy-duty version of the bags from the grocery store. For when it's especially important the bag doesn't rip.

 Hand tools. Be sure you have a hammer, screwdrivers of various sizes, wrenches and pliers, a saw, and so forth so you can take care of emergency repairs.

 Gasoline siphon. A manual siphon only costs a few dollars and can be invaluable if your car or your generator is running out of gas. It's the easiest and safest way to "borrow" gas from one car, for example, to put it in another.

 Big wrench. Be sure you have whatever tool you might need to shut off your water and gas service in case of a leak.

 Yard tools. If you're going to have to clean up, have a strong rake, shovel, and perhaps a pickax.

 Broom. Be sure you have a good broom for the cleanup.

Screws and nails. Have a supply to handle a variety of problems that might come up

 Attic hatch latches or bolts. Get some latches or barrel bolts to hold the attic hatch door in place.

 Utility knife or Swiss Army knife. Don't leave home without it.

Plastic sheeting. Get several good size sheets from the paint department. You'll use these to cover things if you have a leak and to hold water in the tub or sink.

Duct tape. Get a couple of rolls. You can fix almost anything with duct tape.

 Spray paint. If the worst happens, it's the best way to leave a message on your house.

Work gloves. You never know what you'll have to move after the storm.

 Heavy-duty outdoor extension cords. Be sure they're long enough to reach to a neighbor's house, labeled AWG 12, and are brightly colored. If you live in a house, I suggest you get at least two fifty-foot outdoor cords. In an apartment have at least one.

 Power screwdriver. If you're going to be putting up panel or plywood shutters, a good power screwdriver will make it easier.

Waterproof tarp. If you think you might have a roof problem, consider having a waterproof tarp on hand to stop the water damage as soon as possible.

Rope. If you're a homeowner, a good long rope may come in handy. You'll use it to hold trees away from the house or hold the tarp down on the roof, among lots of other situations that can come up.

Wherever-You-Can-Get-It Shopping List

These supplies are available in many locations, and you may have some of them around the house. The important thing is to round them up and keep them with your hurricane supplies.

 Plastic boxes, large and small. Don't scrimp. You'll use these for a million things. They don't have to be expensive, but they need to seal reasonably well. You'll use them to hold hurricane supplies, protect important items, catch water if there are leaks, keep food fresh, and more.

 Water containers. Get collapsible water containers at a store that sells camping supplies and/or big water jugs at the home-improvement store. You'll save money using tap water and have the containers for future storms.

 Fluorescent lanterns. These battery-operated lanterns sit on a table, have a fluorescent bulb, and will run for a long time. They throw light in all directions, so it won't feel so dark without power.

 Flashlights. Everybody in the family needs a flashlight, and you should have a couple of extras.

 Small battery radio. Be sure you have a *small* radio so everybody can listen without headphones during the storm. Consider a larger radio for music and information after the storm, but you'll need lots of batteries.

 Small battery TV. In addition, consider a battery TV. Unfortunately, these days complete information is often not available on the radio after the storm. Most TVs, however, use more batteries than radios do, but small LCD TVs are very efficient.

Car charger for your mobile phone. Charge your phone whenever you're driving.

 Fix-A-Flat. Get a few cans of Fix-A-Flat or a comparable product and keep them in your trunk. Posthurricane debris can be tough on tires.

Crank and Solar-Powered Gadgets

You'll see radios and lights on the market that you crank or shake to generate electricity. I like them only as a backup to the battery-operated devices. You'll go crazy if you have to crank your radio when you want to get information. Besides, small battery radios last a long time. Still, they're not very expensive and a good safety valve for the bottom of your hurricane kit.

I have a little solar-powered radio. It works well as long as it's in the sun, meaning it doesn't work at night.

 Freezer gel packs. Use these in your coolers to keep food cool longer during the storm.

Corded phone. Often the phone service works after the power goes out. Get a cheap phone and a long cord. It should cost about $10.

Extra keys. Store an extra copy of car, work, and house keys with your valuable documents. See page 308.

Extra glasses and contact lenses. Get spares of what you absolutely need and put them with your valuable documents. See page 308.

Battery alarm clock. Be on time even when you don't have electricity.

Fireproof and waterproof lockbox. It's a good idea to always put the important papers you keep at home in a fireproof and water-proof box all year long. If you routinely store valuables there, it will be easy to take them along if you have to leave home. Don't put both keys to the box in the same place.

Coolers. Get several of the cheap styrofoam kind. If the power stays out for a while and you're relying on bags of ice, you'll be glad you did.

Refrigerator thermometer. So you can determine whether your refrigerator stayed cold enough to keep your food from spoiling.

Camping stove or grill. You don't *have* to cook after a storm, but it's nice. Remember, NEVER USE A STOVE OR GRILL INDOORS OR ON AN APARTMENT BALCONY.

Fuel for the stove or grill. Whatever fuel you get, store it in a se-cure place, like the back corner of your garage. Don't store it in-side the house.

Water purification tablets. They usually have them at the drug-store. They're good to have in case you run out of the water you've put aside. Tablets with sodium hypochlorite as the active ingredi-ent are recommended. Iodine-based tablets will work, but could be a problem for some people.

Rabbit ears. Get a cheap antenna with the right connector so you can watch TV if your cable is still out after the electricity comes back on. Test it before the storm to know how to do it later.

Things to Consider Getting—Just to Be Sure

 Pet carrier. If there's any chance you'll have to transport your pet.

 Boots. Anybody who might be working outside after the storm needs heavy work boots that can't be penetrated by nails and debris.

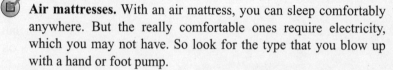 **Sleeping bags.** If you think you might have to evacuate, get one for each person.

Air mattresses. With an air mattress, you can sleep comfortably anywhere. But the really comfortable ones require electricity, which you may not have. So look for the type that you blow up with a hand or foot pump.

Tent. Some people have wanted a tent to sleep in after the storm so they can stay inside the house with a leaky roof.

Flare. Especially if you live in a rural area. After the storm you may need to signal a helicopter so they know where you are.

Acknowledgments

Thanks to St. Martin's Press and my editor, Phil Revzin, you have this second edition of the *Hurricane Almanac* in your hands today. Numerous readers of the first edition wrote in with excellent suggestions. Thanks to all of you. Donna Thomas helped ferret out some of the details you'll find in the history section, and Edward Romero once again ably handled the graphics. I appreciate their help very much.